Lecture Notes in Physics

The Editorial Policy for Proceedings

The series Lecture Notes in Physics reports new developments in physical research and teaching – quickly, informally, and at a high level. The proceedings to be considered for publication in this series should be limited to only a few areas of research, and these should be closely related to each other. The contributions should be of a high standard and should avoid lengthy redraftings of papers already published or about to be published elsewhere. As a whole, the proceedings should aim for a balanced presentation of the theme of the conference including a description of the techniques used and enough motivation for a broad readership. It should not be assumed that the published proceedings must reflect the conference in its entirety. (A listing or abstracts of papers presented at the meeting but not included in the proceedings could be added as an appendix.)

When applying for publication in the series Lecture Notes in Physics the volume's editor(s) should submit sufficient material to enable the series editors and their referees to make a fairly accurate evaluation (e.g. a complete list of speakers and titles of papers to be presented and abstracts). If, based on this information, the proceedings are (tentatively) accepted, the volume's editor(s), whose name(s) will appear on the title pages, should select the papers suitable for publication and have them refereed (as for a journal) when appropriate. As a rule discussions will not be accepted. The series editors and Springer-Verlag will normally not interfere with the detailed editing except in fairly obvious cases or on technical matters.

Final acceptance is expressed by the series editor in charge, in consultation with Springer-Verlag only after receiving the complete manuscript. It might help to send a copy of the authors' manuscripts in advance to the editor in charge to discuss possible revisions with him. As a general rule, the series editor will confirm his tentative acceptance if the final manuscript corresponds to the original concept discussed, if the quality of the contribution meets the requirements of the series, and if the final size of the manuscript does not greatly exceed the number of pages originally agreed upon.

The manuscript should be forwarded to Springer-Verlag shortly after the meeting. In cases of extreme delay (more than six months after the conference) the series editors will check once more the timeliness of the papers. Therefore, the volume's editor(s) should establish strict deadlines, or collect the articles during the conference and have them revised on the spot. If a delay is unavoidable, one should encourage the authors to update their contributions if appropriate. The editors of proceedings are strongly advised to inform contributors about these points at an early stage.

The final manuscript should contain a table of contents and an informative introduction accessible also to readers not particularly familiar with the topic of the conference. The contributions should be in English. The volume's editor(s) should check the contributions for the correct use of language. At Springer-Verlag only the prefaces will be checked by a copy-editor for language and style. Grave linguistic or technical shortcomings may lead to the rejection of contributions by the series editors.

A conference report should not exceed a total of 500 pages. Keeping the size within this bound should be achieved by a stricter selection of articles and not by imposing an upper limit to the length of the individual papers.

Editors receive jointly 30 complimentary copies of their book. They are entitled to purchase further copies of their book at a reduced rate. As a rule no reprints of individual contributions can be supplied. No royalty is paid on Lecture Notes in Physics volumes. Commitment to publish is made by letter of interest rather than by signing a formal contract. Springer-Verlag secures the copyright for each volume.

The Production Process

The books are hardbound, and quality paper appropriate to the needs of the authors is used. Publication time is about ten weeks. More than twenty years of experience guarantee authors the best possible service. To reach the goal of rapid publication at a low price the technique of photographic reproduction from a camera-ready manuscript was chosen. This process shifts the main responsibility for the technical quality considerably from the publisher to the authors. We therefore urge all authors and editors of proceedings to observe very carefully the essentials for the preparation of camera-ready manuscripts, which we will supply on request. This applies especially to the quality of figures and halftones submitted for publication. In addition, it might be useful to look at some of the volumes already published.

As a special service, we offer free of charge LATEX and TEX macro packages to format the text according to Springer-Verlag's quality requirements. We strongly recommend that you make use of this offer, since the result will be a book of considerably improved technical quality.

To avoid mistakes and time-consuming correspondence during the production period the conference editors should request special instructions from the publisher well before the beginning of the conference. Manuscripts not meeting the technical standard of the series will have to be returned for improvement.

For further information please contact Springer-Verlag, Physics Editorial Department V, Tiergartenstrasse 17, W-6900 Heidelberg, FRG

Andrzej A. Zdziarski   Marek Sikora (Eds.)

Relativistic Hadrons in Cosmic Compact Objects

Proceedings of a Workshop
Held in Koninki/Suhora, Poland
9-11 October 1990

Springer-Verlag Berlin Heidelberg GmbH

Editors

Andrzej A. Zdziarski
Marek Sikora
Nicolaus Copernicus Astronomical Center
Bartycka 18, 00-716 Warsaw, Poland

This book was processed by the authors using the T$_E$X macro package from Springer-Verlag

ISBN 978-3-662-13823-6   ISBN 978-3-540-46448-8 (eBook)
DOI 10.1007/978-3-540-46448-8

© Springer-Verlag Berlin Heidelberg 1991
Originally published by Springer-Verlag Berlin Heidelberg New York in 1991
Softcover reprint of the hardcover 1st edition 1991

Typesetting: Camera ready by author

53/3140-543210 - Printed on acid-free paper

Preface

The workshop on *Relativistic Hadrons in Cosmic Compact Objects* took place in the beautiful setting of the Gorce Mountains, near the village of Koninki and Mount Suhora, in the South of Poland, 9–11 October 1990. The drawings shown after the Acknowledgements, by the workshop participant Marina Romanova, give some impression of the beauty and the serenity of the area.

The direct aim of the workshop was to further our understanding of the nature of compact astrophysical sources. The workshop concentrated on energetic hadron processes near compact objects (compact binary systems and active galactic nuclei).

A variety of observational results suggests that relativistic electrons/positrons are produced in the vicinity of these objects and are responsible for radiation spectra extending from the radio to γ-ray regions. The electrons and positrons can be injected into the surrounding area by relativistic hadrons following effective shock acceleration of ions (instead of or in addition to direct acceleration of electrons/positrons). The electromagnetic output is then mediated by such processes as hadron–hadron collisions, photomeson production, and proton–photon pair production. As a result, hadron processes near compact objects can in addition produce large fluxes of very high and ultra-high γ-rays, neutrinos, and free neutrons, all of which may have observable effects.

One of an array of problems discussed at the workshop was the question whether a large fraction of the energy produced by active galactic nuclei is channeled into a population of relativistic protons. It was pointed out by several speakers that reactions involving such protons would lead to an electron–positron pair cascade which could be responsible for the observed nonthermal continuum radiation. Neutrons, which are a byproduct of pion-producing reactions of relativistic protons, can escape from a central region in which they are produced and decay at distances of several light years from the central black hole in an active nucleus. This has dramatic consequences for the dynamics of accretion and the emission of very high energy γ-rays. Direct evidence of relativistic hadron production in active galactic nuclei could come from detection of TeV neutrinos by DUMAND-type detectors.

Another problem that received much attention at the workshop is that of the origin of very high and ultra-high energy γ-rays from galactic compact sources. These sources are believed to consist of close binaries, with gas accreting from a normal star onto the compact object. Here the γ-rays are thought to originate from the impact of relativistic protons on the atmosphere of the normal star. The same suite of hadron processes occurs as in the active galactic nuclei case; the effects of free neutron and neutrino emission remain to be fully explored. By bringing together experts in both the active galactic nuclei and close binary fields, we hope we have stimulated new insights into the behavior of hadrons in these extreme environments.

Warsaw
June 1991

Andrzej A. Zdziarski
Marek Sikora

Acknowledgements

We acknowledge gratefully the financial support from the National Science Foundation, the National Institute of Standards and Technology, and the Polish Academy of Sciences, which support was administered by the Joint Polish-US Maria Curie-Skłodowska Fund. We would like to thank in particular Dr. Ken Gordon of the National Institute of Standards and Technology for his help and encouragement. The support covered a part of the travel expenses and the subsistence of the American and Polish participants. Additional support was provided by the Polish Astronomical Society, to cover a part of the expenses of the young Polish and Eastern European participants.

We would like to thank the other two members of the Scientific Organizing Committee, Professors Mitch Begelman and Julian Krolik, for their efforts in preparing the scientific program of the workshop.

We would like to extend our special thanks to the Local Organizing Committee, which consisted of members of the staff of the N. Copernicus Astronomical Center in Warsaw, and Professor Jerzy Kreiner and his students at the Suhora Astronomical Observatory of the Pedagogical University of Cracow. They handled very smoothly all the logistics of the workshop.

Drawings by the workshop participant Marina Romanova showing the surroundings of the vacation center in Koninki in which the workshop took place.

Contents

List of Participants

Ulrich Achatz	M.P.I. für Radioastronomie, Bonn, Germany
A. Atoyan	Yerevan Physics Institute, Yerevan, Armenia
Stanisław Bajtlik	N. Copernicus Astronomical Center, Warsaw, Poland
Matthew Baring	M.P.I. für Astrophysik, Garching, Germany
Mitch Begelman	JILA, University of Colorado, Boulder, CO, USA
Claes-Ingvar Björnsson	Stockholm Observatory, Saltsjöbaden, Sweden
Gunnlaugur Björnsson	NORDITA, Copenhagen, Denmark
Tzihong Chiueh	I.P.A., National Central University, Taiwan
Paolo Coppi	Caltech, Pasadena, CA, USA
Ocker Cornelis de Jager	Potchefstroom University for C.H.E., Republic of South Africa
Martijn de Kool	JILA, University of Colorado, Boulder, CO, USA
Marek Demiański	Physics Dept., Warsaw University, Warsaw, Poland
Chuck Dermer	Lawrence Livermore National Lab., Livermore, CA, USA
W. I. Dokuchaev	Institute for Nuclear Research, Moscow, USSR
Marek Dróźdź	Physics Institute, WSP, Cracow, Poland
Don Ellison	North Carolina State University, Raleigh, NC, USA
Maria Giler	Physics Institute, University of Łódź, Poland
Peter Giovanoni	NASA, GSFC, Greenbelt, MD, USA
Gilles Henri	M.P.I. für Astrophysik, Garching, Germany
John Kirk	M.P.I. für Kernphysik, Heidelberg, Germany
Jerzy Kreiner	Physics Institute, WSP, Cracow, Poland
Julian Krolik	Johns Hopkins University, Baltimore, MD, USA
Wolfram Krülls	M.P.I. für Radioastronomie, Bonn, Germany
Ewa Kuczawska	Physics Institute, WSP, Cracow, Poland
Zhi-Yun Li	JILA, University of Colorado, CO, USA
R. Mańka	Physics Institute, University of Silesia, Katowice, Poland
Karl Mannheim	M.P.I. für Radioastronomie, Bonn, Germany
Apostolos Mastichiadis	University of Adelaide, Adelaide, Australia
Jeremiah Ostriker	Princeton University Observatory, Princeton, NJ, USA
Michał Ostrowski	Astronomical Observatory, Jagiellonian University, Cracow, Poland
Gabriel Pajdosz	Physics Institute, WSP, Cracow, Poland
Vahe Petrosian	Stanford University, Stanford, CA, USA
Alex Polnarev	Institute for Space Research, Moscow, USSR
Marina Romanova	Institute for Space Research, Moscow, USSR
Maria Sadzińska	Nuclear Research Institute, Łódź, Poland
Marek Sikora	N. Copernicus Astronomical Center, Warsaw, Poland
Maria Soida	Astronomical Observatory, Jagiellonian University, Cracow, Poland

Boris Stern	Institute for Nuclear Research, Moscow, USSR
Roland Svensson	Stockholm Observatory, Saltsjöbaden, Sweden
J. Systka	Physics Institute, University of Silesia, Katowice, Poland
Wiesław Tkaczyk	Physics Institute, University of Lódź, Poland
Giuseppe Vacanti	DPhPE/SEPh, Cen Saclay, Gif/Yvette, France
Gerry Webb	University of Arizona, Tucson, AZ, USA
Magda Zbyszewska	N. Copernicus Astronomical Center, Warsaw, Poland
Andrzej Zdziarski	Space Telescope Science Institute, Baltimore, USA
Janusz Ziółkowski	N. Copernicus Astronomical Center, Warsaw, Poland
Stanisław Zoła	Astronomical Observatory, Jagiellonian University, Cracow, Poland

Relativistic Hadrons Near Accreting Compact Objects: Dynamical and Radiative Signatures

Mitchell C. Begelman

Joint Institute for Laboratory Astrophysics, University of Colorado and
National Institute of Standards and Technology, Boulder, CO 80309-0440
USA

Abstract: Ultrarelativistic protons in the vicinity of an accreting compact object lose energy by interacting with the ambient plasma and radiation field. In the process, they may create significant fluxes of gamma rays, electron–positron pairs, neutrinos and neutrons. I outline the processes by which these particles are created, and their ultimate fates. Pairs and gamma rays participate in an electromagnetic pair cascade, which ultimately contributes to (and may dominate) the nonthermal radiation output from the compact object. Virtually all of the neutrinos and a significant fraction of the neutrons escape, and may be responsible for observable radiative and dynamical effects originating far from the compact object.

1 Introduction

Whether or not ultrarelativistic hadrons are produced in abundance near cosmic compact objects is still an open question. Even assuming that they are produced, their importance in the overall energetics of accreting compact objects is very uncertain. I believe there is a reasonably good chance that the phenomena which form the subject of this meeting do indeed occur. Moreover, as I will argue during my talk, *if* relativistic hadrons are produced efficiently near accreting compact objects, *then* the dynamical and radiative consequences can be quite dramatic, to the extent that we will have to rethink basic aspects of our models for active galactic nuclei (AGN) and X-ray binaries (XRBs). Nevertheless, it will be quite a while before we can make a solid case that relativistic hadrons either are or are not a dominant feature of the environments of accreting compact objects.

One might view the arguments for relativistic hadrons near compact objects as falling into three categories. First, there are arguments based on direct observations of "very-high-energy" (VHE: 0.01–100 TeV) and "ultra-high-energy"

(UHE: > 100 TeV) gamma rays, or other particles, coming from the directions of known XRBs (e.g. Cygnus X-3, Hercules X-1), pulsars (the Crab Nebula) and AGN (Centaurus A). Unfortunately, the only case with reasonably good statistics is the Crab (see Vacanti, this volume), and there we do not know whether the particles are associated with the region very near the pulsar or with the extended nebula. Also worrisome is the fact that different experiments observing the same object have often reached different conclusions (albeit the measurements have generally not been simultaneous). There have been tantalizing reports of periodicities from some XRBs; however, they have not been verified independently. If any of these detections can be shown to be real, it would provide very strong support for the production of ultrarelativistic hadrons, because it is very difficult to accelerate electrons to the required energies. Experimental aspects of this subject should be a central issue for debate at this meeting, and in the community, for some time to come.

I would characterize the second class of arguments as "phenomenological". By this I mean that attempts to explain certain phenomena have led people to suggest that extremely relativistic hadrons are involved. An example of this type of reasoning comes from the modeling of AGN X-ray spectra using highly saturated electron-positron pair cascades (e.g. Zdziarski et al. 1990). To obtain a high degree of saturation it is necessary to start with extremely energetic electrons, which are difficult to accelerate directly. One way to produce UHE electrons with high efficiency is via hadronic reactions (Kazanas and Ellison 1986b; Zdziarski 1986; Sikora et al. 1987).

In the third category, which I call "theoretical", I place the arguments that follow deductively, more or less, from simple assumptions about the nature of accretion flows and the environments of compact objects. These arguments lead to the expectation that the regions close to accreting compact objects are conducive to the production of energetic hadrons. For example, we expect flows near compact objects to be turbulent or at least to involve large concentrations of kinetic energy. It seems reasonable to suppose that there would be shocks near these objects, perhaps very strong (or even relativistic) shocks; theories predict the acceleration of extremely relativistic hadrons in such shocks (see articles by Ellison and Kirk in this volume). Electromagnetic acceleration processes, similar to those that presumably produce jets (Blandford 1989), may also lead to the acceleration of protons to extremely high energies under certain circumstances. If protons are accelerated to ultrarelativistic energies near accreting compact objects, then they must interact with the ambient plasma and radiation field. We follow the consequences of these interactions in Sects. 3 and 4, concluding that a variety of secondary particles are produced which may give rise to observable phenomena. Thus the assumed existence of ultrarelativistic hadrons has well–defined theoretical consequences which are potentially testable.

The conditions near an accreting compact object are very uncertain in detail, but one can make reasonable estimates of important quantities to order of magnitude (Table 1). We are interested in a wide range of masses and scales, ranging from objects of only a few solar masses, like XRBs, up to the black holes in

AGN, which we think have masses of order $10^8 M_\odot$. Lengths and luminosities scale roughly with the mass, and we obtain a correspondingly wide range of radiation densities and magnetic field strengths. However, certain properties of these systems, which happen to be most important for determining the rates of relativistic particle processes, are expected to be insensitive to the mass of the central object. For example, the column densities of particles under both extremes of condition are similar. Since flow velocities are expected to be similar in both situations, the probabilities of particle–particle collisions, e.g. inelastic proton–proton collisions (between an ultrarelativistic proton and a thermal proton in the accretion flow) should also be similar. Factors governing the rates of reactions between protons and ambient photons are more complex, since these reactions tend to occur near threshold (in the center-of-mass frame) and are therefore sensitive to the photon spectrum (see Sect. 3). Note also that XRBs possess a second site where efficient hadronic reactions are possible, in addition to the inner parts of the accretion flow. Interactions of ultrarelativistic particles with the surface of the main sequence secondary star have figured in models to explain the modulation of the UHE flux from objects such as Cygnus X-3 (Vestrand and Eichler 1982; Hillas 1984; Kazanas and Ellison 1986a). To avoid extra complication, I will assume accretion flow conditions in the specific examples that I will present later on.

Table 1. Conditions near accreting compact objects

	XRB	AGN
Mass :	$1 M_\odot$	$10^8 M_\odot$
Luminosity :	10^{38} erg s^{-1}	10^{46} erg s^{-1}
$(\sim L_{Edd})$		
Radius :	10^7 cm	10^{15} cm
$(\sim 30 R_s)$		
Proton		
Density n_p :	10^{17} cm^{-3}	10^9 cm^{-3}
$(v \sim 0.1c)$		
Radiation		
Density :	10^{12} erg cm^{-3}	10^4 erg cm^{-3}
Equipartition		
Mag. Field :	10^7 G	10^3 G
Thomson :		
Opt. Depth :	1	1
$(\tau_T \equiv n_p \sigma_T R)$		

2 Particle Acceleration

Before collisions between extremely relativistic particles can occur, the particles themselves must exist, created presumably by electromagnetic particle acceleration processes. This is a particularly gray area, partly because we know so little about the mechanics of the regions surrounding compact objects and partly because our understanding of particle acceleration is quite limited. For example, we don't know whether the accretion flows are quasi-spherical (Ostriker, this volume), disk-like, or have a more complex structure. The geometry may have a large bearing on whether shocks will form at all, and on the spatial distributions of particles and photons needed for the hadronic reactions to take place. If shocks are present, we don't know whether they would be large coherent structures or a lot of small turbulent structures. The theory of shock acceleration predicts that the energy distribution of accelerated particles depends sensitively on shock strength, speed compared to c, and whether the acceleration occurs all at one shock front or via a series of shock encounters. The observed radiation spectrum, even if it results from nonthermal processes, gives us only indirect clues about the primary hadron spectrum because the radiation is emitted by electrons, not protons. If the regions where particle acceleration occurs contain mainly pair plasma (a possibility if outflowing winds and jets consist mostly of pairs), then there may not be many hadrons to accelerate and the efficiency of relativistic hadron production may be low. On the other hand, if the plasma contains a reasonable admixture of protons, then models for shock acceleration suggest that the acceleration of protons is more efficient than the acceleration of electrons. This is because protons, being more massive than electrons, have larger gyroradii. They see a shock as a much sharper discontinuity than an electron does and gain energy more efficiently through the first-order Fermi process.

The importance of ultrarelativistic hadron interactions will depend not only on the overall efficiency of proton acceleration, but also on the slope of the proton energy distribution and on the maximum energies attained by accelerated protons. If the injection function $S(E)$ is very steep, $d\log S/d\log E < -2$, then most of the energy will be concentrated in particles of low energies. In this case, only a small fraction of the energy will go into particles that can participate in the energetic reactions which I will discuss shortly. On the other hand, it is thought that under conditions that may apply close to a compact object (e.g. relativistic shock speed, very strong shocks), the accelerated particles may develop a rather flat energy distribution through the first-order Fermi process, $d\log S/d\log E \geq -2$. Since the majority of the available energy goes into the most energetic protons in this case, the ultrarelativistic secondary reactions will be much more important. Kirk and Ellison discuss the energy distributions of shock-accelerated protons elsewhere in this volume.

Since the hydromagnetic turbulence that scatters particles in shock accelera- tion is presumably generated by the streaming of the particles themselves, there is no intrinsic scale built into the first-order Fermi process. However, losses will ultimately prevent particles from being accelerated beyond the energy at which the cooling time scale equals the acceleration time. Synchrotron or Compton losses usually limit the energies attainable by electrons; however, these losses probably have little to do with the saturation of proton acceleration. In the absence of effective cooling mechanisms, the acceleration of protons would be limited by the requirement that the gyroradius be smaller than the radius of curvature of the shock. To rough order of magnitude, this would limit proton energies to $\lesssim 3 \times 10^{16}$ eV for XRBs and $\lesssim 3 \times 10^{20}$ eV for AGN. These energies are more than adequate to explain the most energetic particles ($\sim 10^{15}$ eV) yet detected from point sources. Similar energy limits are obtained for electrostatic acceleration along a rotating field line in a charge-separated magnetosphere, as might occur near a pulsar. However, these estimates are overly optimistic, since protons in the environment of an accreting compact object *will* cool, albeit not by synchrotron or Compton processes. As we discuss below, the main energy loss mechanisms at high proton energies ($\gtrsim 10^{13} - 10^{14}$ eV) are proton-photon ($p\gamma$) collisions, which either lead to pair production directly or produce a shower of pions. Assuming the proton acceleration time scale is given by η times the gyro-orbital time, where $\eta \gg 1$, we find that the maximum proton energies are given by

$$E_{\text{max}} \sim 3 \times 10^{14} (\eta/10^4)^{-1/2} \text{ eV} \qquad (1)$$

for XRBs; and

$$E_{\text{max}} \sim 3 \times 10^{16} (\eta/10^4)^{-1/2} \text{ eV} \qquad (2)$$

for AGN, where we have assumed photon energy spectra $F_\nu \propto \nu^{-1}$ and energy densities similar to those given in Table 1. Note that I have used a rather conservative value to normalize η here. These estimates illustrate the tendency for E_{max} to be lower around XRBs than around AGN, suggesting that one might expect to see more energetic particles and photons from the latter. The corresponding maximum energies for electrons, which cool by synchrotron and Compton processes, are six orders of magnitude smaller. Thus, it seems highly unlikely that electrons accelerated directly from the thermal background could be responsible for UHE or even VHE phenomena near compact objects. If such energetic electrons exist, they must be secondary particles created as a result of hadronic reactions.

3 Proton Energy Loss Mechanisms

The principal energy loss mechanisms for ultrarelativistic protons in the central regions of an AGN are discussed in detail by Begelman, Rudak and Sikora (1990; hereafter BRS). As noted above, both synchrotron radiation and Compton scattering by the ultrarelativistic protons are quite unimportant compared to other loss mechanisms, for conditions thought to be present in AGN. The three most important loss mechanisms may be summarized as follows:

- *Inelastic proton-proton collisions.* In this reaction, which is mediated by the strong nuclear force, an ultrarelativistic proton collides with a thermal proton in the background plasma, yielding a pair of ultrarelativistic hadrons of about half the initial total energy plus a shower of pions. The cross section ($\sim 4 \times 10^{-26}$ cm^2) and the inelasticity ($\sim 1/2$) both depend weakly on the energy of the incident proton, E_p, hence the energy loss rate by this process is also insensitive to E_p. However, since the loss rates due to $p\gamma$ processes increase rapidly with E_p (see below), pp reactions are important for cooling protons only below a certain energy. The "grammage" required for energy loss by pp collisions is ~ 80 g cm^{-2}, corresponding to a "Thomson depth" $\tau_T \equiv n_p \sigma_T R \sim 30$. While this may be much larger than the proton column densities thought to exist in most accretion flows, it is important to recall that even the ultrarelativistic protons are dynamically coupled to the thermal plasma by the magnetic field. Therefore, the "effective grammage" in the accretion flow is $\sim 3\tau_T(c/v)$ g cm^{-2}, where v is the characteristic flow speed, which might easily be large enough to lead to cooling. The mean number of pions produced per collision, termed the "multiplicity", is proportional to $\ln E_p$, and is expected to be of order 10–20 for typical energies at which pp collisions dominate proton losses in AGN.

• *Proton–photon pair production.* This is an electromagnetic interaction akin to bremsstrahlung, except that the product is an electron–positron pair instead of a photon. Well above threshold, the product of the inelasticity and cross section is roughly constant, at $\sim 5 \times 10^{-31}$ cm^{-2}. The threshold for this reaction occurs when the photon's energy exceeds $2m_e c^2$ in the proton rest frame. For the maximum proton energies estimated in (1) and (2) (assuming the fiducial value of η), the threshold photon energies are roughly 3 eV and 0.03 eV, respectively; they are of course higher for less energetic protons. These energies are well below the thermal cutoffs expected in the radiation from an optically thick accretion disk (~ 10 keV for an XRB, ~ 100 eV for an AGN). If the disk is responsible for producing *all* of the ambient radiation, then there may be a paucity of low energy photons in the central regions, in which case $p\gamma$ reactions may be quite inefficient compared to pp reactions, even at high proton energies. However, if there is a fairly steep ($d\log F_\nu/d\log \nu \sim -1$) nonthermal radiation spectrum carrying a significant fraction of the radiation output, as there appears to be in AGN, then $p\gamma$ reactions will become the dominant cooling mechanism at high energies, because there will be an increasing number of photons above threshold, as viewed from the proton rest frame. Note that reactions close to threshold will then dominate the cooling.

• *Photomeson production.* In this strong interaction, a proton collides with a photon and produces a hadron plus a shower of pions. At energies far above threshold, the cross section is roughly constant at $\sim 5 \times 10^{-28}$ cm^2, while the inelasticity ranges between ~ 0.14 and ~ 0.5. Although the product of cross section and inelasticity is much larger than that for $p\gamma$ pair production, the threshold for this reaction is much higher: the photon must have an energy exceeding the pion rest mass, ~ 140 MeV, in the proton rest frame. For the maximum proton energies estimated in (1) and (2), the thresholds are roughly 400 eV and 4 eV, respectively, still far below the cutoffs for the thermal disk radiation. For interactions with a nonthermal spectrum, the cooling rate is dominated by reactions near threshold, which implies that the pion multiplicity ~ 1.

Figure 1 shows the relative importance of different proton cooling mechanisms for the simple AGN models studied by BRS. It is assumed that the ambient radiation is dominated by a nonthermal power law with the spectral index $\alpha \equiv -d\log F_\nu/d\log\nu$ varying along the abscissa. The ordinate is the proton Lorentz factor $\gamma_p = E_p/m_p c^2$. Other quantities are held fixed, such as $\tau_T = 1$, total radiation energy density $U_{rad} = 1.2 \times 10^4$ erg cm^{-3}, radius 10^{15} cm, magnetic field strength $B = 10^3$ G, and flow speed $v = 0.1c$. Below the line marked γ_{pp}, inelastic pp collisions are the dominant cooling mechanism, while above it, $p\gamma$ reactions dominate. Because $p\gamma$ pair production has a lower threshold energy than photomeson production, it tends to dominate for steeper nonthermal spectra. Figure 1 shows that the dividing line occurs at $\alpha \approx 1.1$. Note that synchrotron losses become important only for very flat spectra, i.e. where there is a relative paucity of low energy photons. Thus, we might expect a larger role for synchrotron losses if there were no nonthermal component in the ambient radiation spectrum.

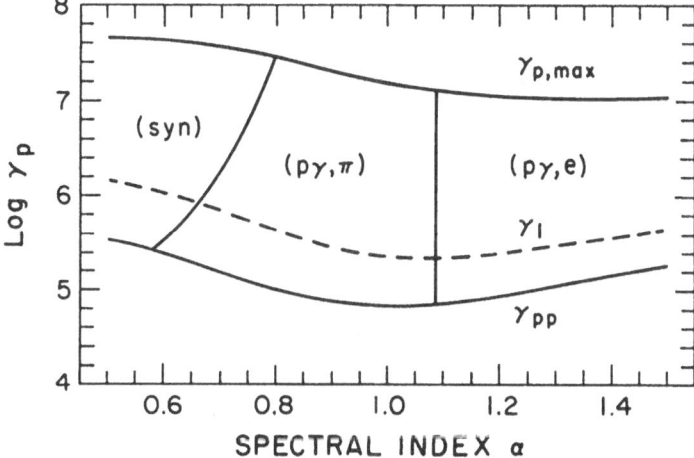

Fig. 1. Relative importance of the main proton cooling mechanisms discussed in the text, as a function of the spectral index of the background radiation and the proton Lorentz factor. (Reproduced from BRS, Fig. 1a.)

The dashed line labeled γ_1 divides the parameter plane according to the efficiency of proton cooling. Below this line, motion of the plasma (at speed v) carries the ultrarelativistic protons away from the central region before they have time to cool. Either they are swept into the black hole (if embedded in inflowing gas) or (if entrained in a wind or jet) they are expelled and suffer adiabatic losses. Above the γ_1 line cooling is complete. Note that γ_1 is 2–3 orders of magnitude smaller than the maximum proton energy, estimated by setting the shock acceleration time equal to the cooling time, with $\eta = 10^4$.

Results for an XRB can be obtained through similar arguments. I reiterate that the role of $p\gamma$ reactions depends sensitively on the photon spectrum, particularly the relative importance of a nonthermal component compared to the thermal disk radiation. Since the maximum energies reached by protons are expected to be ~ 2 orders of magnitude smaller in an XRB than in an AGN while the value of γ_{pp} (assuming a similar nonthermal spectrum) should be unchanged, we might expect $p\gamma$ reactions to be less important in XRBs than in AGNs. The fact that γ_1 may also be similar in the two cases suggests that fewer of the protons will be able to cool fully in the XRB environment.

4 Products of Proton Cooling

Since the protons are confined by the magnetic field and emit little radiation directly, the astrophysically interesting consequences of ultrarelativistic proton production result from the secondary particles which are produced as the protons lose energy. The purely electromagnetic reaction, $p\gamma$ pair production, results in the production of extremely energetic electron–positron pairs. In the radiation environment of the accretion flow, these cool very quickly by producing gamma rays which, in turn, produce additional pairs. Thus, $p\gamma$ pair production will almost certainly lead to an efficient (and perhaps highly saturated) pair cascade. Hadronically induced pair cascades are a straightforward way of obtaining electrons energetic enough to produce UHE gamma rays; they are discussed in more detail by Zdziarski (this volume). The other principal cooling mechanisms, inelastic pp collisions and photomeson production, are basically strong interactions and therefore have a wider range of products. First of all, protons are changed into neutrons in $\sim 25\%$ of the pp reactions and $\sim 50\%$ of the photomeson reactions. Time dilation lengthens the lifetimes of these ultrarelativistic neutrons by many orders of magnitude, and since they are not strongly coupled to the magnetic field they are able to travel large distances before decaying (Biermann and Strittmatter 1987; Sikora, Begelman and Rudak 1989; Kirk and Mastichiadis 1989). Thus, accreting compact objects may be intense sources of free neutrons. The other products of these reactions come from pion decay. Neutral pions decay into a pair of gamma ray photons, which will generally be energetic enough to trigger a pair cascade. Charged pions yield electrons, positrons and neutrinos, via the intermediate stage of muon decay. Table 2 summarizes the branching ratios for particle production by the three main proton cooling mechanisms.

Estimates by BRS show that as much as 10–30% of the energy originally injected in ultrarelativistic protons ultimately goes into neutrons, with a comparable amount going into neutrinos and the remainder shared by pairs and gamma rays. What are the fates of these particles? The pair and gamma ray channels should be considered together, since the cycling back and forth between them is so efficient. Few VHE or UHE gamma rays escape directly from the inner parts of the accretion flow (but see Sikora and Shlosman 1989). Virtually all of the neutrinos will escape, and the detection of these neutrinos would represent the most direct test

Table 2. Products of ultrarelativistic proton cooling

Reaction ($N = n$ or p)	Fraction by Energy			Fraction by Number
	$e^{\pm}$	γ	ν	$n \rightleftharpoons p$
(Np)	1/6	1/3	1/2	1/4
$(p\gamma, e^{\pm})$	1	0	0	0
$(N\gamma, \pi)$	1/8	1/2	3/8	1/2

for ultrarelativistic hadronic processes (see Sec. 5). The fate of neutrons is more complex. Although they are electrically neutral and are not tied to the magnetic field, it is possible for neutrons to suffer np or $n\gamma$ collisions which turn them back into protons before they are able to escape (Sikora, Begelman and Rudak 1989; Kirk and Mastichiadis 1989; BRS).

For a family of simple accretion models parametrized by the Thomson depth $\tau_{\mathrm{T}} = n_{\mathrm{p}}\sigma_{\mathrm{T}}R$, Fig. 2 illustrates the competition of factors that determine neutron production and transport, as a function of neutron energy. In zone A neutrons are produced efficiently, but their escape is hindered by a high probability of reconversion into protons. In zone C it is easy for neutrons to escape but their production is inefficient. Only in zone B are both production and escape of neutrons efficient. These zones translate into an energy spectrum of escaping neutrons which looks quite different from the spectrum of accelerated protons (Fig. 3). The peak in the distribution corresponds to zone B. Since the lifetime of a neutron of Lorentz factor γ_{n} is $\sim 10^{3}\gamma_{\mathrm{n}}$ s, the mean distance traveled before decaying is linearly proportional to the neutron energy, $R_{\mathrm{n}} \sim 10^{-5}\gamma_{\mathrm{n}}$ parsecs. In the case of an AGN, we expect the energy deposition by neutron decay to peak at distances of order 1–10 pc.

5 Observable Consequences

5.1 Direct Observations

The suggestion that hadronic reactions near a compact object might produce a detectable point source of TeV neutrinos goes back to the late 1970's, when water Čerenkov detectors for energetic neutrinos were first being proposed (Eichler 1978, 1979; Berezinsky and Ginzburg 1981). In this type of detector, fast muons created by a reaction involving a muon neutrino produce a cone of Čerenkov radiation, the trajectory and direction of which are measured by an array of photomultiplier tubes. The earliest proposal, DUMAND, consisted of a 1 km^3 cube of water deep in the ocean (Roberts 1976). Recently a number of other detectors have been proposed, e.g. BAIKAL (Allkofer et al. 1990) and GRANDE (Adams et al. 1990a,b; Burman et al. 1990). Although GRANDE is a shallow water detector, it makes up for its lack of volume by using the rock under the water to capture neutrinos coming through the earth. By selecting only upward trajectories, one screens out muons created by gamma rays or hadrons hitting the atmosphere. By using a closely spaced array of photomultipliers it is possible to measure the direction to better than a degree, which means that point sources with count rates of one event per year will give a signal about ten times above the expected background (Berezinsky and Ginzburg 1981). Given that one could detect approximately one in a million neutrinos of energy $\gtrsim 1$ TeV incident of the collecting area of GRANDE ($\sim 6 \times 10^4$ m^2), we conclude that a source with a luminosity in TeV neutrinos of

$$L(\nu_\mu) \gtrsim 6 \times 10^{36} \left(\frac{D}{10 \text{ kpc}}\right)^2 \text{ erg s}^{-1} = 6 \times 10^{44} \left(\frac{D}{100 \text{ Mpc}}\right)^2 \text{ erg s}^{-1} \quad (3)$$

would give about 10 counts per year.

A less direct test for hadronic interactions would be the detection of VHE and UHE gamma rays. As noted earlier, it is difficult to accelerate electrons to such high energies, given the magnetic and radiation densities expected to be present near the compact object. At substantial distances from the central object the limits on electron energy may be relaxed somewhat, but probably will be sufficiently stringent to rule out direct acceleration of electrons capable of producing UHE gamma rays. Unfortunately, VHE or UHE gamma ray photons created close to the compact object will almost certainly pair produce on ambient radiation before escaping. Therefore, observable VHE or UHE gamma rays are likely to be produced mainly in secondary reactions at some distance from the compact object (Kazanas and Ellison 1986b; Sikora and Shlosman 1989; Mastichiadis and Protheroe 1990), most probably in pp reactions. Even at these large distances gamma rays may be prevented from escaping in all directions by radiation from the central source, e.g. escape might be restricted to a small cone around the outward radial direction. Furthermore, pair creation on the microwave background will inhibit the propagation of gamma rays of energy $\gtrsim 10^{15}$ eV over intergalactic distances (Gould and Rephaeli 1978).

5.2 Indirect Observational Signatures

A number of "indirect" observational consequences of hadronic processes have been suggested, some of which will be discussed at this meeting. These include:

- pair cascades triggered by ultrarelativistic pairs and gamma rays produced during the hadron cooling processes. These are discussed by Zdziarski (this volume).

- flat-spectrum parsec-scale radio sources. The electrons released by decaying neutrons will emit synchrotron radiation with a well-defined spatial and spectral signature which depends on the neutron energy distribution function and the radial dependence of the magnetic field. This signature is discussed by Giovanoni and Kazanas (1990; and this volume).

- enhanced boron abundance or other compositional effects due to spallation reactions involving the streaming ultrarelativistic neutrons. These were discussed by Kirk and Mastichiadis (1989). Because the neutrons are so energetic,

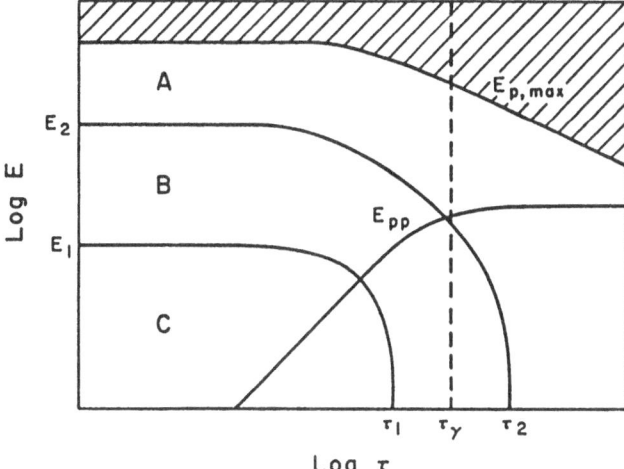

Fig. 2. Characteristic zones for neutron production and escape, as described in text. Inelastic *pp* and *np* collisions dominate below the line labeled "E_{pp}". (Reproduced from BRS, Fig. 4.)

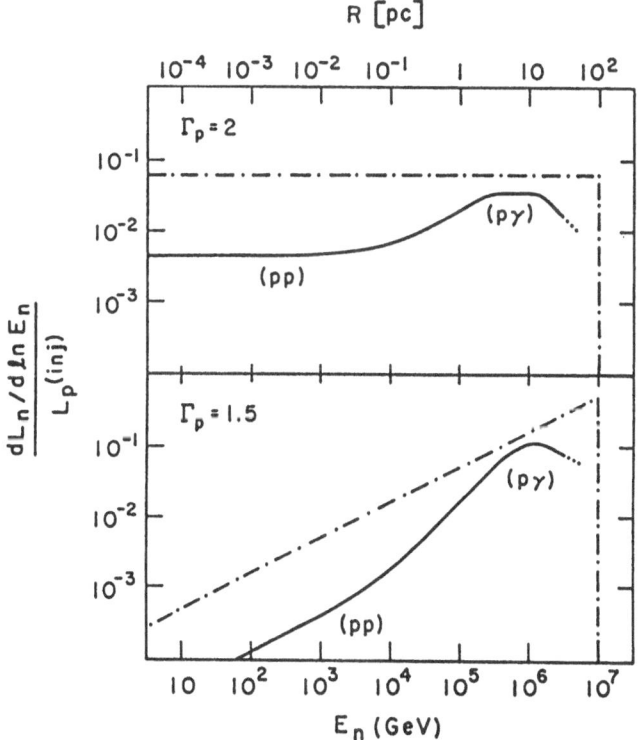

Fig. 3. Energy distribution of escaping neutrons for two proton injection functions with the form $S(E) \propto E^{-\Gamma_p}$. The abscissa is labeled by neutron energy (bottom) and by distance at which neutrons decay (top). For these models, energy deposition by neutron decay peaks at 1–10 pc. (Reproduced from BRS, Fig. 7.)

there are relatively few of them, hence detectable compositional changes will require a long time to develop. According to the estimates by Kirk and Mastichiadis, gas at a few parsecs from an AGN would have to be exposed to the neutron flux for $\gtrsim 10^6$ yr in order to build up a detectable boron abundance anomaly.

- "deep heating" of opaque gas in the broad emission line region of an AGN. Neutrons and extremely energetic gamma rays will be able to penetrate gas to column densities which are completely opaque to X-rays ($N_H \gtrsim 10^{26}$ cm^{-2}). Pairs and gamma rays produced at these depths will ultimately heat the gas, exciting low ionization lines. A further, intriguing possibility is that the ultrarelativistic electrons and positrons will lose energy by synchrotron emission in the EUV and X-ray bands, which can photoionize the gas as well. I am studying these possibilities in collaboration with Ferland and Sikora.

- "limit cycles" and rapid variability may result from the nonlinear couplings in the dynamical reaction networks involving ultrarelativistic hadrons. These are discussed by Stern and Svensson (this volume).

5.3 Dynamical Consequences

Perhaps the dynamical consequences of hadronic reactions are the most interesting. Protons are tied to magnetic field lines and are generally prevented from streaming freely along the field by pitch angle scattering off of Alfvén waves (Wentzel 1974). They are therefore effectively tied to the plasma into which they were injected, and suffer adiabatic losses if they traverse a pressure gradient. Neutrons, on the other hand, stream freely, with *no adiabatic losses*. We have already argued that the bulk of the energy flux carried by streaming neutrons may reach large distances before the neutrons decay, e.g. as much as 1–100 pc in the case of an AGN. Thus a neutron flux offers a virtually loss-free way to transport a large amount of energy from the vicinity of the compact object to a region which may be $10^3 - 10^4$ times further away. When a neutron decays, the resulting proton, which carries 99.9% of the neutron's energy, is trapped locally in the plasma, and its energy is available to drive a wind (through adiabatic expansion) or to heat the thermal gas (through nonlinear damping of Alfvén waves). At such large distances from the compact object, "radiative" losses of the ultrarelativistic protons, i.e. due to pp or $p\gamma$ collisions, are likely to be small, hence virtually all of the energy deposited by the neutrons can be tapped. This form of energy injection provides a way to accelerate a wind to high velocities, starting at large distances from the compact object. Begelman, de Kool and Sikora (1990) have developed a model for the outflows in broad absorption line QSOs which exploits this property to explain one of the longstanding puzzles concerning these objects (see de Kool, this volume).

Surprising as it may seem at first, escaping neutrinos may also produce important dynamical effects. The cross section for a neutrino to interact with a nucleon increases with energy, reaching values $\gtrsim 10^{-35}$ cm^2 at energies $\gtrsim 1$ TeV. As a result, main sequence stars less massive than the sun are opaque to TeV neutrinos.

If the energy flux in neutrinos is sufficiently large, then stars located close to a compact object may absorb a flux of neutrinos that is larger than their intrinsic energy production rate by nuclear reactions. The possibility of heating stars internally by neutrinos was first discussed by Gaisser et al. (1986) for upper main sequence stars in high-mass X-ray binaries. Preliminary stellar structure calculations by M. Czerny, Sikora, de Kool and myself suggest that the bloating of lower main sequence stars by neutrino absorption may be more dramatic, possibly leading to a large increase in radius. This could have important consequences for the ablation of stars close to AGN and for the evaporation of low-mass companions of millisecond pulsars.

This review relies heavily on the long-term contributions of my collaborators: Marek Sikora, Bronek Rudak, Martijn de Kool, John Kirk and Peter Schneider. My work on hadronic processes and compact objects is partially supported by the National Science Foundation, the National Aeronautics and Space Administration, and the Alfred P. Sloan Foundation.

References

Adams, A. et al. 1990a, Proc. 21st ICRC, Adelaide, 2, 439

Adams, A. et al. 1990b, Proc. 21st ICRC, Adelaide, 10, 59

Allkofer, O.C. et al. 1990, Proc. 21st ICRC, Adelaide, 4, 357

Begelman, M.C., de Kool, M., & Sikora, M. 1990, submitted to ApJ

Begelman, M.C., Rudak, B., & Sikora, M. 1990, ApJ, 362, 38 (BRS)

Berezinsky, V.S., & Ginzburg, V.L. 1981, MNRAS, 194, 3

Biermann, P., & Strittmatter, P.A. 1987, ApJ, 322, 643

Blandford, R.D. 1989, in Theory of Accretion Disks, ed. F. Meyer, W.J. Duschl, J. Frank, and E. Meyer-Hofmeister (Kluwer Academic Publishers, Dordrecht), p. 35

Burman, R.L. et al. 1990, Proc. 21st ICRC, Adelaide, 10, 297

Eichler, D. 1978, Nature, 275, 725

Eichler, D. 1979, ApJ, 232, 106

Gaisser, T.K., Stecker, F.W., Harding, A.K., & Barnard, J.J. 1986, ApJ, 309, 674

Giovanoni, P.M., & Kazanas, D. 1990, Nature, 345, 319

Gould, R.J., & Rephaeli, Y. 1978, ApJ, 225, 318

Hillas, A.M. 1984, Nature, 312, 50

Kazanas, D., & Ellison, D.C. 1986a, Nature, 319, 380

Kazanas, D., & Ellison, D.C. 1986b, ApJ, 304, 178

Kirk, J.G., & Mastichiadis, A. 1989, A&A, 211, 75

Mastichiadis, A., & Protheroe, R.J. 1990, MNRAS, 246, 279

Roberts, A., ed. 1976, Proc. 1976 DUMAND Workshop (Fermilab)

Sikora, M., Begelman, M.C., & Rudak, B. 1989, ApJL, 341, L33

Sikora, M., Kirk, J.G., Begelman, M.C., & Schneider, P. 1987, ApJL, 320, L81

Sikora, M., & Shlosman, I. 1989, ApJ, 336, 593

Vestrand, W.T., & Eichler, D. 1982, ApJ, 261, 251

Wentzel, D.G. 1974, AnnRevA&A, 12, 71

Zdziarski, A.A. 1986, ApJ, 305, 45

Zdziarski, A.A., Ghisellini,G., George, I.M., Svensson, R., Fabian, A.C., & Done, C. 1990, ApJL, 363, L1

Can Ultrarelativistic Neutrons from the Central Engine Drive the Outflow in Broad Absorption Line QSOs?

Martijn de Kool [1], Mitchell C. Begelman [1], Marek Sikora [2]

[1] Joint Institute for Laboratory Astrophysics
University of Colorado and NIST, Boulder, CO 80309, USA

[2] N. Copernicus Astronomical Center, Polish Academy of Sciences
Bartycka 18, 00-716 Warsaw, Poland

Abstract: We present a model that can explain many features of the very fast ($\sim 0.1c$) outflows that are observed in broad absorption line QSOs (BALQSOs). In this model, the outflow is accelerated primarily by the pressure of ultrarelativistic protons. These protons are deposited in the wind, over a range of radii, by the decay of ultrarelativistic neutrons escaping from the central engine. Recent models for non-thermal processes in the central engine of an AGN predict such a neutron flux. Since the energy deposition is predicted to peak at distances of order 1-10 pc, the main acceleration will occur outside the broad emission line region (BELR), as is required by the observations.

The absorbing gas occupies a very small fraction fraction of the volume of the wind, and must be confined by it. The most likely confining mechanism is thermal pressure of the wind. The ionization parameter of the absorbing gas has to be sufficiently low for it to remain cool, which requires that the wind is very efficiently heated by the ultrarelativistic protons. This could be accomplished through the excitation and subsequent damping of Alfvén waves, or by a turbulent wind structure containing shocks that will drive the thermal and cosmic ray energy density to equipartition.

A simple model in which the absorption is caused by a fine spray of cloudlets, comoving with the wind and in pressure equilibrium with it, can reproduce a wide variety of BAL profiles similar to those that are observed. Line profiles prove to be much more sensitive to ionization effects than to the dynamics of the wind, thus making it difficult to derive the dynamical properties of the wind from the line profiles.

1 Introduction

The properties of broad absorption line QSOs have been most recently reviewed by Turnshek (1988; hereafter T88) and Weymann, Turnshek and Christiansen (1985; hereafter WTC). The defining characteristics of these objects are the presence of strong, very broad (velocities up to 0.2 c) absorption in the ultraviolet resonance lines like those of C IV, Si IV and N V, which is blueshifted relative to the center of the broad emission line. A study of the properties of these absorption lines leads to the conclusion that they have to be caused by a fine spray of cold clouds (size $\sim 10^{11}$ cm) with a very low ($\sim 10^{-6}$) volume filling factor. One of the outstanding problems in the interpretation of these lines is that they appear to be formed outside the BELR. The evidence for this is that in some cases the entire blue wing of the BEL is absorbed, and in other cases the Lyman α BEL is absorbed by the N V BAL. This implies that there is material with low velocity outside the BEL, and it is likely that the outflow which causes the BALs is accelerated outside the BELR. Another unexplained problem concerning BALs is how the absorbing material is kept at a high enough gas pressure to remain cool in the radiation field of the AGN, or in other words what confines the absorbing clouds.

In a number of recent papers (Sikora, Begelman and Rudak 1989, Kirk and Mastichiadis 1989, Begelman Rudak and Sikora 1990; herafter BRS) on the consequences of high energy processes occuring in the central engine of an AGN, it has been shown that a significant part of the luminosity generated in such objects could escape in the form of a flux of relativistic neutrons. These neutrons are created by inelastic proton-proton collisions and photomeson reactions involving protons accelerated up to energies as large as $10^7 - 10^8$ GeV. Because of relativistic time dilation neutrons with Lorentz factor γ_n can travel a distance $\sim 10^{-5}\gamma_n$ pc before decaying. The relativistic protons resulting from neutron decay are trapped by the magnetic field, and their energy may be used for bulk acceleration and heating of the ambient gas. BRS estimate that most of the energy will be deposited between 1 and 10 pc, and point out that this energy deposition could easily drive an outflow with the main acceleration occurring in the radius range required for BALQSOs. In this paper we report on the results of a closer investigation of this idea, in which we are constructing a detailed model of the BALR to see if the observed BALs can actually be obtained in such a scenario.

2 The Dynamics of Neutron Driven Winds

The energy distribution of the relativistic neutrons escaping from the central engine is discussed by BRS. The neutrons are produced as a result of collisions of very energetic protons (Lorentz factor γ_p up to $10^7 - 10^8$), accelerated in shocks in the central engine, with ambient photons and thermal protons. Proton-photon ($p\gamma$) collisions are the dominant production mechanism for high energy neutrons, whereas proton- proton collisions are responsible for the low-energy neutrons. The emerging neutron spectrum can be divided into several regimes, according to the relative importance of the physical processes shaping the spectrum. These processes are: i) neutron escape; ii)photomeson reactions ($p\gamma$ and $n\gamma$ collisions); iii) inelastic pp and np collisions and iv) removal from the central engine by accretion. The relative importance of these processes varies with energy, and in each energy range the neutron spectrum can be estimated by taking only the dominant processes into account. For the conditions expected in the central engine of an AGN, and assuming a primary proton injection function (by number) which is a power law in energy ($Q_p \propto \gamma_p^{-\Gamma_p}$), the neutron spectrum is found to be given by:

$$\frac{dL_n}{d\ln\gamma_n} \propto \gamma_n^{-\Gamma_n}$$

with Γ_n changing from $\Gamma_p - 2$ to $\Gamma_p - 3$ to $\Gamma_p - 2$ to $\Gamma_p - 1$ as the neutron energy increases. The positions of the breaks in this spectrum that signify the transitions between the different physical regimes depend on the specific parameters of the central engine, and the reader is referred to BRS for a detailed discussion. For QSO-like sources with a sufficiently flat proton injection function ($\Gamma_p \leq 2$), the bulk of the neutron luminosity will be emitted around $\gamma_n \sim 10^5 - 10^6$. Because of relativistic time dilation these neutrons will decay at a distance of order 1-10 pc from the central engine.

We assume that neutrons escaping from the central engine travel ballistically until they decay, at which point the resulting protons are trapped by the magnetic field. The energy deposition rate per unit volume in relativistic protons ('cosmic rays' or 'CR') at radius R is therefore given by

$$H_{CR}(R) = \left[\frac{dL_n/d\ln\gamma_n}{4\pi R^3}\right]_{\gamma_n = R/c\tau_n} \equiv \frac{L'_{CR}(R)}{4\pi R^3}$$

where $\tau_n \simeq 10^3$ s is the mean neutron lifetime in its rest frame and the neutron energy spectrum $dL_n/d\ln\gamma_n$ is described above.

The ultrarelativistic protons lose energy mainly by adiabatic expansion, which converts CR energy into kinetic energy, and through the excitation of damped Alfvén waves, which converts CR energy into thermal energy of the background plasma. The latter process occurs at a rate (per unit volume) of

$$C_{CR}^A = -v_{CR}\frac{dp_{CR}}{dr}$$

where v_{CR} is the cosmic ray streaming speed relative to the plasma. v_{CR} is likely to be of the order of the radial component of the Alfvén speed v_A (Skilling 1975), but its value can not yet be accurately predicted.

Using the prescriptions for energy injection and coupling between CR and thermal plasma described above, we have constructed some simple spherically symmetric hydrodynamical models of neutron driven winds with constant mass flux. We use a two fluid approximation for the thermal background plasma and the cosmic rays, because the absorbing clouds are transparent to CR, and the ionization parameter of the absorbing material is determined by the pressure of the thermal background plasma. Therefore, this quantity has to be calculated even if it is not important for the dynamics of the wind. Because of the uncertainties in the value of v_{CR}, we considered three different cases : case a) has no coupling between CR and thermal plasma at all, case b) has extremely efficient coupling, so that the injected CR energy is instantaneously converted into heat, and case c) is an intermediate case, in which v_{CR} is given by a fixed fraction of the wind speed v. Radiative energy losses of the thermal plasma are always taken into account.

The results of these models are the following:

1) Neutron injection is quite effective in driving a wind at large radii. The largest velocities are reached at scales of order 100 pc, i.e. well outside the region between 1 and 10 pc where most of the energy is injected. The acceleration process is very efficient in the sense that most injected energy is converted into kinetic energy of the wind. Even in the case that all injected energy is immediately converted into heat only 20% of L_n goes into Compton losses, and 80 % ends up as wind kinetic energy.

2) Wind velocity profiles are very smooth and do not differ too much between the different cases. Therefore, it will be impossible to derive any constraint on the dynamics from the BAL profiles.

3) Case b) is the only model in which the thermal pressure can easily provide confinement for the BAL clouds. Confinement is marginally possible in case c), but only if the ratio between v_{CR} and v is greater than 0.5.

4) Both case b) and case c) have the property that the ionization parameter is essentially constant over the acceleration region, as seems to be required by a comparison of the line profiles of the different ions.

3 Broad Absorption Line Profiles

To test the neutron driven wind model further, we have attempted to calculate actual BAL profiles that would be expected in our model, again using a number of simplifying assumptions: the absorbing clouds move with the wind; the clouds have equal mass; the cloud temperature is $10^{4.5}$ K; the clouds are in pressure equilibrium with the thermal pressure in the wind. We have used the dynamical model derived from case b) in this section, since in this case strong BALs are most easily produced. Because the clouds are much smaller than the background source, both geometrical covering and covering in velocity space have to be taken into account when deriving the line profile. We use the formalism developed for this situation by Kwan (1990). For a given dynamical model, the line profile can be calculated if we specify the the mass loss rate in cold material, the mass of one cloud, the intrinsic Doppler width of one cloud, and the ionizing spectrum from the central source. The latter is assumed to consist of the sum of an optical to soft X-ray power law with a spectral index –1.4, an X-ray power law with index –0.7 and a black body component with a temperature of 5×10^4 K. The power laws are assumed to be equal at 1 keV, and the thermal bump contains the same power as the optical to X-ray power law. When calculating the ionization equilibrium in the clouds as a function of radius, this spectrum is modified by absorption due to HI, HeI and HeII in the interior part of the outflow.

Reasonable line profiles are obtained for a mass loss rate in cold material of $\sim 0.15 M_\odot/\mathrm{yr}$, which is only a few percent of the mass flux required in hot material to provide sufficient confinement for the clouds, which is of order 10 $M_\odot/\mathrm{yr}$ (all assuming spherically symmetric outflow). This justifies our neglect of the clouds in the dynamical models.

The result we find is that the line profiles for a given mass loss rate are influenced mainly by ionization and geometrical covering effects. Ionization levels in clouds at a certain velocity can change drastically for slightly different prescriptions of how the cold material is loaded into the wind at small radii, mainly because of the resulting changes in the internal absorption. The strongest lines are found for an intermediate amount of internal absorption, i.e when the ionizing EUV radiation is significantly reduced but not completely gone. When the cold material is concentrated in clouds that are too massive, geometrical covering becomes incomplete and the BAL becomes less strong. For the cold mass loss rate quoted above, the cloud mass where this starts to happen is around 10^{17} gr, and the BAL disappears completely for cloud masses greater than 10^{20} gr.

Even for the smooth distribution of clouds over velocity and space that we use to calculate line profiles, structure can appear in the BALs due to ionization and geometrical covering effects.

Properties of observed BALs that are impossible to reproduce in our model are a lack of absorption at low velocity and the presence of multiple distinct sharp absorption components. However, all these observations can be accomodated in the model in an obvious way if the unrealistically simple assumptions of a steady, spherically symmetric wind containing equal mass absorbing clouds are dropped.

4 Summary and Discussion

We have shown how the escape of ultrarelativistic neutrons from the central engine of an AGN can lead to the production of a strong, fast wind with most of the acceleration occurring outside the broad emission line region. This feature is required in order to explain observations of BALQSOs, in which the blue wings of the BELs are often absorbed. A fine spray of absorbing cloudlets, accelerated along with the wind yields BAL profiles similar to those which are observed. In order to maintain a sufficiently low level of ionization in the cloudlets to produce the absorption, it is necessary that a large fraction of the relativistic proton energy be converted to thermal pressure, which confines the cloudlets. Photoelectric absorption of portions of the AGN continuum spectrum, possibly by gas in the inner parts of the wind, makes it easier to obtain the relatively low ionization levels observed.

The production of a powerful neutron flux, containing up to 10% or more of the bolometric luminosity of the AGN, is inevitable if the nonthermal acceleration of protons to ultrarelativistic energies is energetically important in the central engine. Although the typical AGN continuum spectrum has been succesfully modeled on the assumption that these processes take place (Zdziarski *et al.* 1990), there is no strong evidence that such acceleration is efficient, nor is it known whether the injected proton energy distribution is sufficiently flat or extends to high e-nough energies ($\sim 10^5$ GeV or greater) for our model to be viable. Direct tests of the ultrarelativistic proton hypothesis would involve the detection of neutri-nos with energies greater than 1 TeV (BRS) or VHE gamma rays (Mastichiadis and Protheroe 1990), and will not be feasible for several years at least. Besides reflecting the primary injection function of protons, the energy distribution of es-caping neutrons (and hence the radial distribution of energy injection in the wind) is shaped by the efficiency of neutron production (at low neutron energies) and the trapping of neutrons in the central engine by photomeson collisions (at high energies). These effects depend in turn on the detailed gas dynamics and radiative transfer within the central engine, which are not well understood. However, our conclusion (from BRS) that the energy deposition function should peak roughly in the range 1–10 pc should not be too sensitive to these details.

For simplicity, we calculated spherically symmetric wind models. Although the neutron flux is likely to be nearly isotropic, there are strong observational arguments that the BAL region covers a solid angle of only $\sim$ 1 sr. Turnshek (1988) considers whether this partial covering is confined to a single region of the sphere, e.g., a cone or a disk, or consists of random patches. The fact that many cloudlets must lie along each line of sight within the BAL zone argues for a well-defined region, as does the presence of multiple troughs in many sources. In our model, the partial covering of the BAL zone presumably results from the angular distribution of sources of absorbing gas, which must be entrained by the wind. Outside the solid angle which contains the absorbing gas, we predict the existence of a fast (possibly relativistic), hot wind, but with a density so low that it may be virtually undetectable by direct observation. Since the wind carries such a large energy flux, however, it may be detectable through its interaction with the interstellar medium of the host galaxy or with the surrounding intergalactic medium, at distances well beyond the BAL region.

References

Begelman, M.C., Rudak, B., and Sikora, M. 1990, *Ap. J.*, **362**, 38 (BRS).

Kirk, J., and Mastichiadis, A. 1989, *Astr. Ap*, **211**, 75.

Kwan, J. 1990, *Ap. J.*, **353**, 123.

Mastichiadis, A., and Protheroe, R.J. 1990, *M. N. R. A. S.*, **246**, 279.

Sikora, M., Begelman, M.C., and Rudak, B. 1989, *Ap. J. Lett.*, **341**, L33.

Skilling, J. 1975, *M. N. R. A. S.*, **172**, 551.

Turnshek, D.A., 1988, in *QSO Absorption Lines: Probing the Universe*, S. C. Blades, D. A. Turnshek and C. A. Norman (Editors), Space Telescope Science Institute Symposium 2, Cambridge University Press, Cambridge, U.K., p. 17 (T88).

Weymann, R.J., Turnshek, D.A., and Christiansen, W.A., 1985, in *Astrophysics of Active Galaxies and Quasi-Stellar Objects*, J. Miller (Editor), Oxford University Press, Oxford, U.K., p. 333 (WTC).

Zdziarski, A.A., Ghisellini, G., George, I.M., Svensson, R., Fabian, A.C., and Done, C. 1990, *Ap. J. Lett.*, **363**, L1.

Inflation of Stars by TeV Neutrinos

Michał Czerny [1], Marek Sikora [2], Mitchell C. Begelman [3]

[1]Space Research Centre, Warszawa, Poland
[2]N. Copernicus Astronomical Centre, Warszawa, Poland
[3]JILA, Boulder, University of Colorado and NIST, Boulder, USA

Abstract: Irradiation by the mixture of TeV neutrinos and electromagnetic radiation has a strong impact on the structure of low-mass stars. The radii of such stars may be increased by a few orders of magnitude. This effect may have very important consequences for processes in the vicinity of supermassive black holes in Active Galactic Nuclei and some X-ray binary systems.

Active Galactic Nuclei (AGN) are extremely strong sources of electromagnetic radiation. Luminosities of some AGN may exceed 10^{47} erg s^{-1}. Recently Begelman, Rudak and Sikora (1990) proposed that AGN also produce strong fluxes of cosmic radiation as well as very hard neutrinos. The neutrino luminosity may be of order 10% of the electromagnetic luminosity. The cross-section for absorption of such energetic neutrinos (typical energies $10 - 100$ TeV) is so large that stars may become opaque for this kind of radiation. Therefore hard neutrinos can be an additional source of heating and strongly influence stellar structure. Besides AGN, the effect of neutrino heating may also be important in close binaries, as shown by Gaisser *et al.* (1986).

Tout *et al.* (1989) investigated the impact of electromagnetic radiation on the structure of stars placed in the proximity of AGN. They put a star in a bath of radiation of brightness temperature T_b. The energy density of this bath is $\sigma T_b^4/\pi c$. The bath temperature T_b may be related to the luminosity of the AGN by the formula

$$log\ T_b\ =\ 3.792\ +\ 0.25\ log\ L_{47}\ -\ 0.5\ log\ d_{0.1},$$

where L_{47} is the luminosity in units of 10^{47} erg s^{-1}, and $d_{0.1}$ is the distance from the AGN in units of 0.1 pc. As the electromagnetic radiation is absorbed in the very outer layers of the star, Tout *et al.* computed evolutionary stellar models in the usual way, changing only the outer boundary conditions. They found that the electromagnetic radiation has a rather small effect on main sequence stars, but can strongly increase radii and mass loss rates from giants.

We compute the structure of a 0.3 $M_\odot$ main sequence star exposed to both electromagnetic radiation and TeV neutrinos. The treatment of electromagnetic radiation is similar to that of Tout *et al.* Additionally we assume that the bath

contains neutrinos. The energy density of neutrinos is 10 per cent of the electromagnetic energy density. For simplicity we assume that neutrinos are monochro-
matic. We take the cross–section for neutrino absorption to be 10^{-34} cm^2, which
corresponds to a neutrino energy of about 30 TeV (Stecker 1979).

We employ the following numerical scheme. We compute the model of a star
not exposed to any kind of radiation (*i.e.* a classical main sequence star). Knowing
the density profile we can compute what fraction of neutrino flux is absorbed by
each stellar layer. The absorption gives rise to electromagnetic stellar flux. We
compute the profile of this flux and add it to the flux produced by thermonuclear
reactions. We also change the outer boundary conditions due to the effects of the
electromagnetic bath. As the new model has a different density profile than the
starting one, we repeat the computations until stellar parameters do not change.
We then increase the energy density of the bath and iterate the next model.

The stellar radius as a function of the temperature of the bath is presented in
Figure 1 (solid line). Clearly there are two branches. The lower branch (labelled
"optically thick") corresponds to large optical depths for neutrino absorption,
τ_ν, exceeding 30. (τ_ν decreases monotonically along the line from bottom left to
top right.) At the bottom of the branch labelled "optically thin" τ_ν is about 10,
decreasing to about 0.03 for the last computed model. The radii of stars on the
optically thin branch are much bigger than those on the other branch. They are
a strong function of T_b, and may exceed 10 $R_\odot$. From Figure 1 one can see that
these effects are caused mainly by TeV neutrinos, as changes in radius caused by
the electromagnetic radiation only (dashed line) are much smaller.

Besides radius, other stellar parameters change significantly as well. Moving
along the solid line in Figure 1 from bottom left to top right, the luminosity due to
nuclear reactions decreases very rapidly. On the optically thin branch it becomes
completely negligible, and practically the whole electromagnetic flux produced
within the star is due to absorption of neutrinos. Also the central temperature
and the central density decrease. For the last computed model their values are
only 1.8×10^5 K and 1.2×10^{-3} g cm^{-3}, respectively. Therefore stars on the
optically thin branch do not resemble ordinary giants (although they also have
large radii), since they do not exhibit dense cores. Rather, their temperature and
density profiles flatten.

Our results show that irradiation by TeV neutrinos may have a very significant
impact on low mass main sequence stars. In particular, they markedly increase
the stellar radius and correspondingly decrease the surface gravity. As a result,
mass loss due to X-ray heating, as well as the rate of stellar collisions, may be
strongly enhanced close to an AGN. Mass released during these processes may
become an important source of fuel for the supermassive black hole, and/or supply
mass to a wind. Emission lines observed from AGN may be produced in released
matter and/or atmospheres of inflated stars. Neutrino irradiation of low mass
stars might also be important in X-ray binaries such as Cygnus X-3 and in close
binaries containing a millisecond pulsar, and could conceivably contribute to the
evaporation of the companion in the latter class of objects.

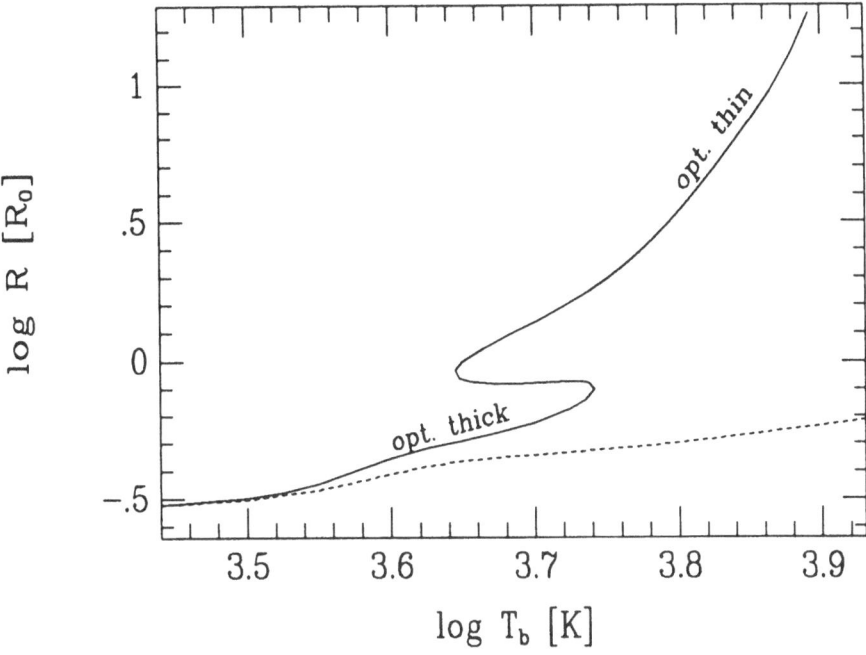

Fig. 1. Equilibrium radii of a 0.3 $M_\odot$ main sequence star immersed in a bath of radiation and TeV neutrinos. The absissa indicates the brightness temperature of the incident radiation. The solid line corresponds to a neutrino density equal to 10% of the radiation density, while the dashed line shows the response of the star when neutrinos are absent. The star is opaque to the neutrinos on the branch labeled "optically thick" and transparent on the branch labeled "optically thin".

References

Begelman, M.C., Rudak, B. and Sikora, M., 1990, *Ap.J.*, **362**, 38.
Gaisser, T.K., Stecker, F.W., Harding, A.K. and Barnard, J.J., 1986, *Ap.J.*, **309**, 674.
Stecker, F.W., 1979, *Ap.J.*, **228**, 919.
Tout, C.A., Eggleton, P.P., Fabian, A.C. and Pringle, J.E., 1989, *M.N.R.A.S.*, **238**, 427.

Electron-Positron Pair Production in Compact Sources with Energetic Protons

Andrzej A. Zdziarski

N. Copernicus Astronomical Center
Bartycka 18, PL-00-716 Warsaw, Poland
and
Space Telescope Science Institute
3700 San Martin Dr., Baltimore, MD 21218, USA

Abstract: Selected theoretical models of X-ray and γ-ray compact sources in which the available power is supplied to either thermal or nonthermal protons, and in which electron-positron pair production occurs are discussed. The pair production is due to electron and photon interactions in the ambient thermal electron plasma Coulomb-heated to a relativistic temperature and/or electromagnetic cascades initiated by decay of pions produced by energetic protons. The models assume either spherical or disk accretion onto a black hole or approximate the source as a one-zone region. The steady state pair abundance and instabilities related to pairs are discussed.

1 Introduction

Energization of protons (and α particles) much more effective than that of electrons is often expected in the vicinities of compact objects. This process can then give rise to production of either thermal or nonthermal electron pairs, analogously to the situation in which electrons are energized directly. The energetic protons either heat up ambient thermal electrons to relativistic temperatures or produce pions in proton-proton and proton-photon collisions and $e^{\pm}$ pairs in proton-photon collisions. Then further $e^{\pm}$ pairs are produced by electron-electron, electron-photon, and photon-photon interactions (e.g., Svensson 1984). The pions decay either into hard γ-rays, which subsequently produce pairs in collisions with soft photons, or into positrons or electrons. These kinds of proton-induced pair production are the subject of this contribution.

In a thermal scenario, protons accrete onto a compact object (e.g., a black hole), in which process they acquire a subvirial temperature, $kT_p \lesssim 0.1 G m_p / r$. Then, the thermal protons transfer some of their energy to the electrons via Coulomb interactions (and possibly plasma collective interactions). The electrons can be kept in this way at a relativistic temperature, at which copious electron-positron pairs are produced. Also, protons at a near-virial temperature can produce pions, which decay products give then rise to an electron-positron pair cascade.

In a nonthermal scenario, protons are accelerated to relativistic energies, e.g., in the first-order Fermi process, occurring in shocks. The protons, as heavier particles, are again accelerated much more effectively than electrons, with the expected flux of protons about 10^2 times higher than that of electrons at the same energy according to some calculations (Bell 1978). The relativistic protons can produce directly $e^\pm$ pairs in proton-photon collisions, and indirectly via pion production in proton-photon and proton-proton collisions. The maximum number of $e^\pm$ pairs per proton is about 5, which follows from the accretion efficiency $\lesssim 0.1$ and the effectiveness of the conversion of nonthermal luminosity into the rest mass of pairs $\lesssim 0.1$ (Svensson 1987, 1990).

In this contribution, we assume that the reader is familiar with both thermal and nonthermal pair production in compact sources. For a current review, see, e.g., Svensson (1990). We also refer the reader to Svensson (1990) for a review of pairs in accretion flows from an angle different than that adopted in this contribution.

The major parameter determining the amount of pair production in a source of size R and luminosity L is its compactness,

$$\ell \equiv \frac{L}{R} \frac{\sigma_T}{m_e c^3} = \frac{2\pi}{3} \frac{m_p}{m_e} \left(\frac{L}{L_E}\right)\left(\frac{3R_S}{R}\right). \tag{1}$$

Here, σ_T is the Thomson cross section, L_E is the Eddington luminosity, and $R_S \equiv 2GM/c^2$ is the Schwarzschild radius. If $L = L_E$ and $R = 3R_S$, $\ell \simeq 3600$. If the luminosity L is emitted mostly in the soft γ-ray range, the source becomes optically thick to pair production in photon-photon collisions when $\ell \gtrsim 10$ (Guilbert, Fabian, and Rees 1983).

2 Spherical Black Hole Accretion

Spherical accretion can be hydrodynamic or collisionless. In the former case, one assumes that accreting protons interact effectively with each other, with the mean free path much shorter than the distance to the black hole, and behave like fluid, accreting on a time scale of the order of the free fall time scale. The short mean free path can be due to, e.g., magnetic fields irregular on small scales. In the latter case, proton interactions are ineffective, and the protons orbit the black hole in a star-like manner. In that case, the accretion time scale is determined by the rate of the proton energy loss and may be much longer than the free fall time scale.

The accretion efficiency, e ($\equiv L/\dot{M}c^2$, where $\dot{M}$ is the mass accretion rate), is in some models self-consistently solved for, based on the values of the velocity of the accretion flow and the effectiveness of assumed radiative processes. This efficiency is often very small, $e \ll 0.1$ (e.g., Shapiro and Teukolsky 1983). On the other hand, it has been postulated (Mészáros 1975) that dissipative processes related to the presence of chaotic magnetic fields can greatly increase the efficiency. Typically, it is assumed in those dissipative accretion models that a fixed fraction ($\sim 1/2$) of the gravitational energy available at $R \gtrsim 3R_S$ is converted into the electron thermal energy, which is then radiated away. This leads to accretion efficiencies of ~ 0.1.

2.1 Thermal Adiabatic Hydrodynamic Accretion

Steady-state adiabatic hydrodynamic spherical accretion including effects of $e^{\pm}$ pairs has recently been studied by Park and Ostriker (1989) and Park (1990a, b). Their models incorporate relativistic hydrodynamics, radiation moment equations, bremsstrahlung and Comptonization as the main radiative process, and take into account the possible presence of a shock. The models are self-consistent in the sense that the accretion efficiency is solved for and that effects of radiation produced at one radius on the accretion flow at other radii (preheating) are included. The accretion is adiabatic in the sense that only the adiabatic heating of the electrons and protons occurs. It is assumed that magnetic fields are sufficient to fluidize accretion flows, but small enough not to give rise to any substantial synchrotron radiation.

Park (1990a) considers one-temperature accretion, i.e., one with the electron temperature, T_e, equal to the proton temperature, T_p. Except for highly supercritical accretion rates, this assumption requires that a process much more effective than Coulomb scattering couples the protons and electrons (e.g., Begelman and Chiueh 1988). Park finds single solutions with high temperatures (the maximum $T \sim 10^9$–10^{10} K) for $\dot{m} < 0.1$, where $\dot{m} \equiv \dot{M}c^2/L_E$. At $\dot{m} \geq 0.1$, the temperature of the flow drops to 10^4. In addition, a second high-temperature solution appears at $3 \lesssim \dot{m} \lesssim 100$ and is characterized by relatively high efficiency, $e \sim 10^{-4}$, and $L/L_E \sim 10^{-3}$. He finds that all the solutions have negligible pair density. The low-temperature flow is too cool to produce any pairs. The high-temperature one at $\dot{m} \lesssim 30$ has radiation temperatures sufficient for pair production but too low compactness. On the other hand, models with $\dot{m} \gtrsim 30$ have enough photons but insufficient radiation temperatures.

Park (1990b) considers two-temperature accretion. He finds single self-consistent solutions with high temperatures (the maximum $T_p \sim 10^{11}$ K, $T_e \sim 10^{10}K$) at $\dot{m} \lesssim 0.1$ and $5 \lesssim \dot{m} \lesssim 100$. Those solutions are very close to the high-temperature solutions of one-temperature accretion. In fact, at $\dot{m} \gtrsim 30$, electrons and ions become fully coupled, and the both cases become identical. He gives the maximum relative pair density in these solutions as < 0.03.

A second type of solutions for two-temperature accretion is found by Park and Ostriker (1989). They find high-temperature solutions with abundant $e^{\pm}$ pairs for $3 \lesssim \dot{m} \lesssim 40$. The pair density is up to a few times that of ionization electrons. The efficiencies and luminosities are relatively high, $e \sim 10^{-2}$ and $L/L_E \sim 10^{-2}$. The two different states of flow at a given $\dot{m}$ can be regarded as two different pair equilibrium states of relativistic thermal plasma (Lightman 1982; Svensson 1982, 1984).

One should note, however, that the high pair-density solutions are not obtained by integration from the accretion radius, in contrast to the low pair-density solutions. Rather, they are obtained assuming the density and (high) temperature at $10^3 R_S$. Unlike the low pair-density models, they cannot be connected to the steady-state outer part of the flow and require suppressing in some way the pre-heating disruption (Park 1990b). Park and Ostriker (1989) argue that any related instability will have a time scale much longer than the dynamical time scale at $10^3 R_S$ and the inner part of the flow can be considered time-independent.

Park (1990b) considers high-energy γ-ray production from pions produced by hot thermal ions in two-temperature flows. He finds the γ-ray luminosity to be a significant fraction of the total luminosity only for $\dot{m} \lesssim 10^{-3}$, which corresponds only to very weak sources. Thus, pion production in hydrodynamic thermal accretion appears not to be a process relevant to cosmic γ-ray sources. This is due to the flow temperature being too low for efficient pion production to proceed.

The treatment of pair processes in Park and Ostriker (1989) and Park (1990b) is rather simplified; for example, a constant cross section above the threshold for photon-photon pair production is assumed. A more refined treatment may change the quantitative results of those papers.

Stability of the models of Park and Ostriker (1989) and Park (1990a, b) has not been investigated and it is possible that those models would show flaring instability on both short and long time scales.

Only the models with abundant pairs can correspond to observable cosmic sources, such as active galactic nuclei (AGNs). The other models predict luminosities many orders of magnitude below the Eddington limit, which is very unlikely for AGNs and galactic X-ray sources. The radiation temperature in the pair-abundant models can be as high as 1 MeV, which can in principle account for the observed X-rays and γ-rays. The spectra from accretion flows in those models are essentially those of Comptonized bremsstrahlung, which tends to have the spectral index less than zero. This is much too hard for typical AGNs with the average X-ray energy spectral index of $\alpha \simeq 0.7$ (e.g., Turner and Pounds 1989; the spectral index is defined here by $F_E \propto E^{-\alpha}$).

2.2 Dissipative Hydrodynamic Accretion

Thermal models of this class have been dealt with by, e.g., Mészáros (1975), and Maraschi et al. (1979, 1982). Those authors argue that the available gravitational energy can be efficiently converted into heat by turbulence and magnetic dissipation. For a Schwarzschild black hole, the efficiency can be then as high as $e \sim 0.05$–0.1. The magnetic field is assumed to be in equipartition with the gravitational energy and $T_e = T_p$ is assumed. The thermal cyclo-synchrotron, bremsstrahlung and multiple Compton scattering processes are then fast enough to radiate away almost all the thermal energy. The infall velocity is less than the free fall velocity by a factor of ~ 2, to account for the large fraction of the gravitational energy converted into heat. The adiabatic heating is only about $\sim 5\%$ of the dissipative turbulent heating. The possible disruption of the flow by preheating is not investigated (cf. Park 1990a).

The maximum temperatures of the flow of $T_e \sim 10^9$–10^{10} K at the accretion rates $2 < \dot{m} < 20$ are obtained in the model. This is quite similar to the one-temperature non-dissipative models of Park (1990a). This coincidence is due to the near cancellation of the change in the heating rate, which is now higher due to dissipation, with the change in the cooling rate, which is now also higher due to the presence of equipartition magnetic fields. In contrast to the model of Park (1990a), a near-isothermal high temperature zone at $R \lesssim 10^2 R_S$ appears in the present models. The spectra at infinity are nearly power laws, and are due to multiple Compton scattering of the self-absorbed cyclo-synchrotron radiation. The power laws extend from the infrared to soft γ-rays. The typical spectral indices obtained by Maraschi $et\ al.$ (1982) are $\alpha \sim 0.5$-1, in agreement with the typical X-ray spectra of AGNs. The obtained spectra are, however, in disagreement with the AGN IR and UV spectra, which are more complex than simple power laws.

Electron-positron pairs are not included in those models. Pairs will be present, however, in the conditions of $kT_e \sim m_e c^2$ and $\tau_T \sim 1$, where τ_T is the Thomson optical depth (e.g., Svensson 1984; Zdziarski 1985), characteristic of this class of models. As a near-isothermal inner zone appears in the model, one can use the pair-equilibrium study of Zdziarski (1985) to infer the average temperature and pair abundance. Such calculations are done in Zdziarski (1988). He finds solutions with abundant pairs (the positron density, $n_+ \lesssim 5n_i$, the ion density) for $0.2 \lesssim \dot{m} \lesssim 5$ at the inflow velocity of 1/2 of the free fall. An effect of pairs is to lower the temperature and increase the optical depth. The emitted spectra are now harder (of $\alpha \sim 0$) but cut off at lower energies. The total luminosity is not affected by the pairs, as the dissipation of gravitational energy is related here to turbulent and magnetic processes and its efficiency is assumed $a\ priori$. No instability due to pairs is expected. The pairs reduce the Eddington luminosity, but they also cause a reduction in the actual luminosity by increasing the trapping radius in the spherical flow (Zdziarski 1988).

Begelman, Sikora, and Rees (1987) and Tritz and Tsuruta (1989) have considered two-temperature dissipative quasi-spherical accretion flows with pair production (without integrating the flow dynamic equations). Tritz and Tsuruta (1989) have assumed the presence of external photons and no magnetic fields. Begelman $et\ al.$ (1987) and Tritz and Tsuruta (1989) find a critical accretion rate, $\dot{m}_{crit} \sim 1$, above which the hot protons will cool within some radius on the inflow time scale due to Coulomb interactions with ionization electrons and $e^{\pm}$ pairs. Then, the flow is expected to collapse to a geometrically thin accretion disk. This effect would occur at the lower $\dot{m}_{crit}$ and larger R if pairs were present. However, Tritz and Tsuruta (1989) have found only few pairs in these two-temperature accretion flows if the pairs are produced only thermally.

Tritz and Tsuruta (1989) and Lightman, Zdziarski and Rees (1987) have consider spherical dissipative accretion flows with nonthermal pair production. The nonthermal pair production is treated in a generic manner, using an approximation to the rate of conversion of nonthermal energy into the rest mass of pairs by Lightman and Zdziarski (1987). Tritz and Tsuruta (1989) have found that abundant pairs may be present in this case in two-temperature flows. They then affect the collapse of the ion-supported flows. Lightman $et\ al.$ (1987) have found abundant pairs ($n_+ \lesssim 10n_i$) for $0.03 \lesssim \dot{m} \lesssim 30$. The pairs lower then the Eddington luminosity and increase the trapping radius, similarly to the thermal case discussed (Zdziarski 1988).

2.3 Thermal Collisionless Accretion

In this class of models, one typically assumes that the accretion flow becomes collisionless downstream a standing shock, which appears at some $R \gg R_S$ (e.g., Mészáros and Ostriker 1983). This effect is due to heating the downstream protons to a near-virial temperature, at which they become collisionless provided magnetic fields are sufficiently small. The magnetic field must be many orders of magnitude below its equipartition value. Mészáros and Ostriker (1983) argue that downstream magnetic fields decay fast via conductivity.

The downstream protons orbit the black hole in a star-like manner. The accretion rate is then equal to the time scale of proton energy loss due to Coulomb collisions with electrons. The electron temperature is typically much less than the proton temperature. As the time scale for proton energy loss may be much longer than that of the free fall, the obtained efficiencies can be also much larger, $e \sim 0.1$.

The heated electrons can achieve relativistic temperatures. They radiate via bremsstrahlung, synchrotron, Compton and neutral-pion decay processes, and the spectra can extend to the X-ray and γ-ray bands. Mészáros and Ostriker (1983) have not calculated any effects due to the presence of $e^{\pm}$ pair production in their model.

Those effects have been considered by Moskalik and Sikora (1986). They consider a time-dependent source at a constant proton temperature and a constant matter inflow rate from external region. The proton advection rate equals their cooling rate. The rate of change of the $e^{\pm}$ pair density equals the difference between the pair production rate due to photon-photon, photon-particle, and particle-particle interactions and the pair annihilation rate (pair advection and escape were neglected). They also consider the rate of change of the photon number and the electron energy density. They find that there is a narrow range of the values of the proton inflow rate at which there is no stable equilibrium of the source. For the case they consider with $kT_p = 60m_e c^2$ (of the order of the proton virial temperature at $R = 10R_S$), the unstable range corresponds to the average luminosity close to $10^{-2}L_E(10R_S/R)$, or, equivalently, to $\ell \simeq 10$. Note that the proton temperature (corresponding roughly to the position of the shock) is not determined self-consistent here, but assumed, which may cause the absence of the instability in a realistic accretion flow model (cf. Babul, Ostriker and Mészáros 1989; see below).

In the unstable range, the source exhibits a limit cycle behavior with recurrent flares. The flare timescale equals approximately the diffusion time, $\tau_T R/c$, and the recurrence period is found to be about $10^2 R/c$. During the recurrence period, the energy density in the protons is gradually accumulated, and during the flash, there is a pair production runaway and the accumulated energy gets transferred to $e^{\pm}$ pairs and, almost instantaneously, to their radiation.

Babul *et al.* (1989) have extended the accretion model of Mészáros and Ostriker (1983), including the effects of pairs in a very approximate manner. They consider supercritical accretion rates, for which solutions with a shock are found. The shock exists typically at $R \sim 10^2 R_S$. Downstream the shock, there exists a collisionless transition region of the kind discussed in Mészáros and Ostriker (1983). The thickness of the transition region is assumed to equal the mean free path for proton cooling. The pair density in the transition region is typically 20% of the proton density, and the luminosity is $\sim 0.1L_E$. This luminosity and $R \sim 10^2 R_S$ correspond to the same compactness, $\ell \sim 10$, at which the flaring instability was found by Moskalik and Sikora (1986). However, the typical proton temperatures found by Babul et al. (1989) are $\sim 5m_e c^2$ (of the order of the virial temperature at $10^2 R_S$), much less than the value of $60m_e c^2$ assumed by Moskalik and Sikora (1986).

2.4 Nonthermal Collisionless Accretion

This class of models is similar to that discussed above in §2.3 except that a mechanism producing nonthermal, relativistic particles is invoked. This mechanism can be the first order Fermi acceleration in an accretion shock (Protheroe and Kazanas 1983; Kazanas and Ellison 1986a). The shock is supported by the pressure of collisionless nonthermal protons downstream the shock. The magnetic field is assumed to be strong enough to confine the protons in the shock region, but small enough not to fluidize them. This requirements roughly correspond to the gyroradius of the most energetic protons being smaller than the shock radius but larger than the Schwarzshild radius. The latter condition corresponds to the rather high maximum Lorentz factor of the accelerated protons $\gamma \gtrsim 10^8 (B/1G)(M/10^9 M_\odot)$.

The relativistic protons lose energy through pion production in collisions with ambient protons, provided the accretion rate is large enough. This contrasts the case of protons in thermal accretion flow, which lose energy mostly via Coulomb interactions with ambient electrons. The balance between the relativistic proton energy loss rate and the injection rate determines then the energy density downstream and the position of the shock (Kazanas and Ellison 1986a). The final products of pion decay are γ-rays, electrons and positrons, and neutrinos. Their production rates are calculated by Protheroe and Kazanas (1983). The primary γ-rays and $e^\pm$ pairs give rise to an $e^\pm$ cascade, which is the process determining the form of radiated photons. The X-ray and γ-ray spectra from this process are qualitatively similar to those of AGNs and the cosmic γ-ray background (Kazanas and Protheroe 1983).

3 Disk Accretion

3.1 Radiation-Pressure Dominated Disks

The "standard disk model" is that of Shakura and Sunyaev (1973). Their model assumes that the disk is geometrically thin and that the rate of angular momentum transfer is proportional to the total pressure times the constant viscosity parameter α_v, with $\alpha_v \lesssim 1$ typically. The pressure in an inner region of the disk is typically dominated by radiation and the proton and electron temperatures are equal. When the disk is effectively optically thick in the vertical direction, its local radiation field can be approximated as modified blackbody. Then, the maximum temperature of a disk is always nonrelativistic, for both stellar mass as well as supermassive black holes, and there is no $e^{\pm}$ pair production.

However, the innermost region of the disk is effectively optically thin, although the scattering optical depth is still very large. Then the modified blackbody approximation is no longer valid. Rather, one needs to explicitly consider the radiation production mechanisms, with bremsstrahlung and Comptonization being the major processes. Maraschi and Molendi (1990) have studied the temperatures and radiation spectra assuming that the inner part of the disk is homogeneous and isothermal in the vertical direction. They find that if $\dot{m} \sim 1$ and $\alpha_v \sim 1$, the maximum disk temperatures can be as high as $kT \sim 100$ keV. Maraschi and Molendi (1990) do not include effects of $e^{\pm}$ pair production although they notice that they would be present at this temperature. One of the effects of pairs would be to limit the temperature, as pair production increases the medium density (e.g., Svensson 1984).

Maraschi and Molendi (1990) find that the temperatures of $kT \gtrsim 10$ keV, which are necessary for this model to account for the AGN X-ray spectra, are obtained only for large accretion rates, $\dot{m} \gtrsim 0.5$ and for $\alpha_v \sim 1$. However, when $\dot{m}$ becomes comparable to unity, the assumption of geometrical thinness breaks down. The effects of the increase in disk thickness on the temperature and spectra are not investigated in that paper. Thus, it is not clear whether the results of the paper are valid for $0.5 \lesssim \dot{m} \lesssim 1$, and whether $kT \gtrsim 10$ keV can be actually obtained in radiation-pressure dominated disks. Even if this is so, the required narrow range of $\dot{m}$ limits the applicability of this model to AGNs, which probably span a much larger range of $\dot{m}$ (e.g., Padovani and Rafanelli 1988).

3.2 Gas-Pressure Dominated Disks

A solution to disk accretion in which the conditions for the presence of copious $e^{\pm}$ pairs are satisfied is that of gas-pressure dominated effectively optically thin disk of Shapiro, Lightman, and Eardley (1976). In most models, protons and electrons are assumed to be thermal but may have different temperatures, $T_p \sim 10^{11}$ K, and $T_e \sim 10^9$ K. As in dissipative thermal spherical accretion (§2.2), the protons are heated by some dissipative processes and cooled through Coulomb collisions with the electrons. The electrons radiate away the received energy.

The effects of electron-positron pairs in such disks have been studied by Kusunose and Takahara (1988, 1989), Tritz and Tsuruta (1989), White and Lightman (1989, 1990), and Björnsson and Svensson (1991b, c). White and Lightman (1989) have considered several kinds of steady-state thermal accretion disks. They find $e^{\pm}$ pairs may be important in two-temperature Comptonized bremsstrahlung disks. For the accretion rates $\dot{m} > \dot{m}_{crit} \lesssim 1$, there are no pair-equilibrium solutions in an annulus centered on $R \simeq 5R_S$. In the stable regions of R and $\dot{m}$, there are two solutions, one with high pair density, and one with the pair density much less than the proton one. The high pair density appears unphysical, as $T_p > T_{virial}$ in a large range of R, as well as unstable to pair runaway (White and Lightman 1990). The low pair density solution is stable (White and Lightman 1990). If the electrons and protons are effectively coupled and $T_e = T_p$, the critical accretion rate is lower and the electron temperatures are higher (White and Lightman 1989).

The disk at radii at which no pair-equilibrium solutions exist is likely to collapse to an optically thick, radiation-pressure dominated configuration (White and Lightman 1990).

Two-temperature disks with a supply of external soft photons have been considered by White and Lightman (1989) and Tritz and Tsuruta (1989). They find that solutions with unsaturated Comptonization of soft photons (with $y = 1$, where $y \sim [kT_e/m_ec^2][\tau_T + \tau_T^2]$ is the Compton parameter) are stable to pair runaway and have negligible pair density. No gas-pressure dominated solutions are found for the case where an effective proton-electron coupling results in $T_e = T_p$ (White and Lightman 1989).

Björnsson and Svensson (1991b) have shown that the equations describing effectively thin accretion disks can be rewritten as two relations between the accretion disk parameters, $\dot{m}$, the viscosity parameter, α_v, and the radius, R, and the parameters of pair-equilibrium sources, the *local* compactness, ℓ, and the local optical depth of ionization electrons, τ_p. The results of pair equilibrium studies can be then directly mapped onto the disk parameter space. Each solution in the pair plasma studies then corresponds to a specific accretion rate, the viscosity parameter, α_v, and the radius, R. This constitutes a set of generic disk solutions for given viscosity parameters. Björnsson and Svensson (1991b) applied their method to the case of disks where the only source of photons is optically thin bremsstrahlung.

Björnsson and Svensson (1991c) give some preliminary results for the case of Comptonized bremsstrahlung disks. That specific disk solution can be obtained from the generic solution of Björnsson and Svensson (1981b). Their findings differ at some points from those of White and Lightman (1989). It is found that pairs become important for $\dot{m} \gtrsim 1$. For $\alpha_v < \alpha_{crit} \sim 1$, infinite radial gradients appear when pairs become important and the standard disk assumptions break down. For $\alpha_v > \alpha_{crit}$, fully self-consistent disk solutions with the pair density of the order of the ion density exist.

Tritz and Tsuruta (1989) have considered also accretion disks with nonthermal pair production and Comptonization of external soft photons (with the Compton parameter $y = 1$). They find high equilibrium pair densities for $\alpha_v \sim 1$ and $\dot{m} \sim 1$. Some effects of the pairs are to inflate the disk, lower the ion density, increase the ion temperature, and decrease the electron temperature.

4 One-Zone Models

In this section, selected models that treat the source as a single homogeneous and isotropic zone are discussed. Most of those models invoke accretion in some way.

4.1 Thermal Models

Sikora and Zbyszewska (1985) consider an inner, quasi-symmetric, zone of a two-temperature black hole accretion flow assuming steady state, Coulomb heating of electrons by protons kept at a constant temperature, $e^{\pm}$ pair equilibrium, and Comptonized bremsstrahlung as the dominant radiative process. The last assumption requires that any magnetic field in the source is of the strength much below equipartition. They find that such a source can in principle account for the soft γ-ray spectra observed from strong γ-ray emitting AGNs (e.g., NGC 4151, MCG 8-11-11). However, a rather low compactness for the innermost region of the accretion flow, $\ell \lesssim 10$, is required. An effective shielding from the UV radiation emitted outside the γ-ray source (e.g., from an accretion disk) is required as well.

The most current study of stability of thermal plasma clouds with electrons and pairs heated by protons is that of Björnsson and Svensson (1991a; see there for earlier references). They perform a linear stability analysis for a source with a constant power supplied to the protons but with the constant density of protons (in contrast, e.g., to the study of Moskalik and Sikora 1986) and neglecting pair advection/outflow. They include energy conservation for both electrons and protons (rather than to assume a constant T_p). They find a rather small instability region, with $10 < \ell < 25$ and the optical depth of ionization electrons $\tau_p < 2$.

Dermer (1988) considers a magnetized thermal two-temperature source, with the magnetic field energy density in equipartition with the radiation energy density. The proton temperature is taken as a free parameter. The electron temperature follows from the balance between Coulomb heating and radiative losses. He considers production of positively charged pions in collisions of thermal protons. Then positrons from pion decay radiate soft photons in the synchrotron process. Those soft photons are then Comptonized by thermal electrons, forming a power law spectrum extending to the X-ray regime. He finds that the presence of the soft photons from pion-decay positrons has a stabilizing effect on the spectral index of the Comptonization spectrum. The spectral index turns out to be close to $\alpha = 0.7$, which is the average value observed in hard X-rays from AGNs (e.g., Turner and Pounds 1989). Dermer (1988) does not give the obtained values of the obtained electron temperatures. Thus it is difficult to assess the importance of thermal $e^{\pm}$ pair production in the model.

Some effects are not taken into account in the model. Irradiation of the source by the UV bump radiation, prominent in AGNs, would provide additional electron cooling and soften the X-ray spectra. Also, γ-rays from neutral pion decay would be pair-absorbed in typical AGN conditions. The resulting pairs would outnumber the positrons from the charged pion decay and produce more synchrotron soft photons than included in the model. This would have a further cooling and softening effect.

4.2 Nonthermal Models

Zdziarski (1986) considers a nonthermal source in which protons are accelerated to relativistic energies at the steep rate of $\dot{N}(\gamma) \propto \gamma^{-\Gamma}$, $\Gamma = 2.4$. This power law index is expected from Fermi acceleration in moderately strong shocks and it it similar to that inferred for the source of galactic cosmic rays (e.g., Bell 1978). The process of pion production in collisions of the relativistic protons with the thermal protons of the accretion flow is considered. This process gives rise to nonthermal power law electrons and positrons generated at a power law rate with the same index as that of the primary protons, $\Gamma = 2.4$. There is a kinematic low-energy cutoff in the production rate at $\gamma_{\min} \sim 100\text{-}300$. The electrons and positrons cool in the synchrotron and Compton processes. The magnetic field energy density is taken to be in equipartition with the radiation energy density. The cooling process gives rise to a broken power-law steady-state electron distribution, $N(\gamma) \propto \gamma^{-p}$, with $p = 2$ and 3.4 below and above $\gamma_{\min}$, respectively. This broken power law electron distribution gives rise to a broad-band photon spectrum similar to that of radio-quiet AGNs. The radio and the far IR spectra are synchrotron self-absorbed. The near IR and optical spectra are dominated by synchrotron radiation of $p = 3.4$ electrons and have the spectral index of $\alpha = 1.2$. The UV bump is assumed to come from another source and it provides only additional seed photons for Comptonization. The hard X-ray spectrum is due to Comptonization by both $p = 2$ and 3.4 electrons and its spectral index turns out to be $\alpha \sim 0.7\text{-}0.9$, in a qualitative agreement with the AGN observations.

The model cannot, however, reproduce spectra with $\alpha \lesssim 0.7$, which are quite common in AGNs (e.g., Turner and Pounds 1989). Also, the model predicts that the IR variability time scale should be qualitatively similar to that in X-rays, which has not been observed.

Sikora et al. (1987) consider acceleration of protons to relativistic energies at a flat power law rate, with $\Gamma < 2$. They point out the importance of proton-photon collisions for the proton cooling and production of $e^{\pm}$ pairs and γ-rays. One possible process is photo-pion production, $p\gamma \to p(n)\pi$, in which a relativistic proton produces neutral or charged pions in collisions with soft photons. If a positively charged pion is produced, then the proton changes its state into a neutron. The other process is production of an $e^{\pm}$ pair, $p\gamma \to pe^{+}e^{-}$. Proton-photon cooling dominates cooling via inelastic proton-proton collisions at high enough energies. For example, for $\dot{m} = 0.1$ and a power law spectrum of soft photons with $\alpha = 1$, the proton-photon processes dominate at Lorentz factors $\gamma \gtrsim 10^5$. If the soft photon spectrum is a power law, the photo-pion production dominates over proton-photon pair production for $\alpha \lesssim 1$. The both processes give rise to production of nonthermal relativistic $e^{\pm}$ pairs with power law spectra with two different normalizations and cut off energies but with the spectral index same as that of the accelerated protons.

Kazanas and Ellison (1986b), Sikora, Begelman and Rudak (1989), Kirk and Mastichiadis (1989), and Atoyan (1990) point out the importance of generation of relativistic neutrons in the proton-proton and proton-photon collisions. The decay time of neutrons in the source frame is about $10^3\gamma$ s. In contrast to protons, the neutrons are not scattered by chaotic magnetic fields and can travel ballistically. The maximum Lorentz factor of accelerated protons is determined by the balance of the acceleration rate in a shock with the cooling rate and can be as high at $\gamma_{max} \sim 10^8$ (Sikora *et al.* 1987). The maximum Lorentz factors of escaping neutrons is somewhat less ($\sim 10^7$), as they have to be able to escape the central source without undergoing a pion-production reaction (Sikora *et al.* 1989). Thus, the escaping neutrons can travel as far as 100 pc before decaying. Upon decay, they transfer their energy to the ambient matter, which can, e.g., generate large scale winds and inhibit accretion (Sikora *et al.* 1989, Begelman, Rudak and Sikora 1990). The neutrons escaping to large distances in AGNs can also give rise to a detectable flux of VHE and UHE γ-rays from decay of pions produced in hadronic interactions (Mastichiadis and Protheroe 1990).

Begelman *et al.* (1990) consider in some detail hadronic-electromagnetic cascades generated by proton acceleration, taking into account the neutron production. They consider an example in which the dominant electron cooling process is synchrotron radiation. In the cascades, protons produce pions and $e^{\pm}$ pairs, the pions decay into γ-rays and electrons and positrons, the electrons and positrons radiate synchrotron photons and Compton upscatter soft photons, and γ-rays produce further $e^{\pm}$ pairs.

Stern, Svensson and Sikora (1990) and Stern and Svensson (1991) consider an isolated compact source with constant power used to inject monoenergetic relativistic protons. The injected power gets converted into soft radiation by an electromagnetic cascade. If the Lorentz factor of the injected protons is low enough, the source becomes unstable and developes limit cycles with the proton and the photon components interchangeably dominating (somewhat analogously to the model of Moskalik and Sikora 1986, in which the $e^{\pm}$ pair and the proton components interchangeably dominated). This may cause periodic large amplitude variations of the luminosity from a compact object. A possibly stabilizing effect of external radiation on the nonthermal source has not been investigated.

5 Summary

Electron-positron pair production in compact sources in which protons receive most of the available power has been reviewed. The theoretical models have been divided into categories based on the source geometry (spherically symmetric accretion, disk accretion, "generic" one-zone models) and on the distribution of energetic protons (thermal or nonthermal).

Pairs are found important in several models. A high pair density solution is found for hydrodynamic adiabatic spherical accretion (Park and Ostriker 1989). The solution may be unstable on long time scales due to preheating.

Abundant pairs are present in quasi-spherical accretion flows with turbulent dissipation and either thermal or nonthermal pair production for some flow parameters (Begelman *et al.* 1987; Lightman *et al.* 1987; Zdziarski 1988; Tritz and Tsuruta 1989). An effect of pairs is then to reduce both the Eddington luminosity and the actual luminosity (due to an increase in the trapping radius; Lightman *et al.* 1987; Zdziarski 1988). Another effect is to enhance a rapid cooling of protons in the inner parts of the flow, which is likely to lead to a collapse of the quasi-spherical flow to an optically thick geometically thin disk (Tritz and Tsuruta 1989).

Pairs may be abundant and lead to a limit cycle instability in thermal collisionless spherical accretion with $T_p \gg T_e$ and with a shock (Moskalik and Sikora 1986). However, the study of Babul *et al.* (1989) indicates that the shock may occur at a rather large radius ($R \sim 10^2 R_S$), at which the proton temperature is too low to initiate efficient pair production and the instability.

Pairs are unimportant in radiation-pressure dominated accretion disks, except for $\dot{m} \sim 1$ and $\alpha_v \sim 1$ (Maraschi and Molendi 1990). They are important and may lead to instabities in hot gas-pressure dominated Comptonized bremsstrahlung disks (White and Lightman 1989, 1990; Björnsson and Svensson 1991c). Pair densities are negligible in disks with a copious external source of soft photons (e.g., from the outer regions of the disk; White and Lightman 1989; Tritz and Tsuruta 1989). Pairs are found important in soft-photon cooled disks with nonthermal pair production (Tritz and Tsuruta 1989).

The presence of abundant pairs is predicted in models treating the source as a one-zone region in which protons are energized provided the source compactness, ℓ, is large enough and the cooling is not too efficient (see, e.g., Björnsson and Svensson 1991a for a thermal model; Stern and Svensson 1991 for a nonthermal model). The standard theories of thermal and nonthermal pair equilibrium (e.g., Svensson 1984, 1987) apply then with some modifications. The presence of pairs may lead to instabilities in both thermal (Björnsson and Svensson 1991a) and nonthermal (Stern and Svensson 1991) cases. The instabilities are likely to disappear in the presence of copious external soft photon source.

In general, high pair densitites are more likely in nonthermal models than in thermal models. The former *assume* efficient acceleration of particles, which can then easily initiate pair production. This contrasts thermal models, in which the electron distribution is exponentially cutoff at high energies and the temperature adjusts itself to the heating, cooling, and pair production processes. This adjustment has then a limiting effect on the rate of pair production.

I would like to thank Marek Sikora and Gabriele Ghisellini for valuable discussions.

References

Atoyan, A. M. 1990, preprint.
Babul, A., Ostriker, J. P., and Mészáros, P. 1989, *Ap. J.*, **347**, 59.
Begelman, M. C., and Chiueh, T. 1988, *Ap. J.*, **332**, 872.
Begelman, M. C., Rudak, B., and Sikora, M. 1990, *Ap. J.*, **362**, 38.
Begelman, M. C., Sikora, M., and Rees, M. J. 1987, *Ap. J.*, **313**, 689.
Bell, A. R. 1978, *M. N. R. A. S.*, **182**, 443.
Björnsson, G., and Svensson, R. 1991a, *M. N. R. A. S.*, **249**, 177.

Björnsson, G., and Svensson, R. 1991b, *Ap. J. (Letters)*, **371**, L69.
Björnsson, G., and Svensson, R. 1991c, this volume.
Dermer, C. D. 1988, *Ap. J. (Letters)*, **335**, L5.
Guilbert, P. W., Fabian, A. C., and Rees, M. 1983, *M. N. R. A. S.*, **205**, 593.
Kazanas, D., and Ellison, D. C. 1986a, *Ap. J.*, **304**, 178.
Kazanas, D., and Ellison, D. C. 1986b, *Nature*, **319**, 380.
Kazanas, D., and Protheroe, R. J. 1983, *Nature*, **302**, 228.
Kirk, J. G., and Mastichiadis, A. 1989, *Astr. Ap.*, **213**, 75.
Kusunose, M., and Takahara, F. 1988, *Publ. Astr. Soc. Japan*, **40**, 435.
Kusunose, M., and Takahara, F. 1989, *Publ. Astr. Soc. Japan*, **41**, 263.
Lightman, A. P. 1982, *Ap. J.*, **253**, 842.
Lightman, A. P., and Zdziarski, A. A. 1987, *Ap. J.*, **319**, 643.
Lightman, A. P., Zdziarski, A. A., and Rees, M. J. 1987, *Ap. J. (Letters)*, **315**, L113.
Maraschi, L., and Molendi, S. 1990, *Ap. J.*, **353**, 452.
Maraschi, L., Perola, G. C., Reina, C., and Treves, A. 1979, *Ap. J.*, **230**, 243.
Maraschi, L., Roasio, R., and Treves, A. 1982, *Ap. J.*, **253**, 312.
Mastichiadis, A., and Protheroe, R. J. 1990, *Astr. Ap.*, **246**, 279.
Mészáros, P. 1975, *Astr. Ap.*, **44**, 59.
Mészáros, P., and Ostriker, J. P. 1983, *Ap. J. (Letters)*, **273**, L59.
Moskalik, P., and Sikora, M. 1986, *Nature*, **319**, 649.
Padovani, P., and Rafanelli, P. 1988, *Astr. Ap.*, **205**, 53.
Park, M.-G. 1990a, *Ap. J.*, **354**, 64.
Park, M.-G. 1990b, *Ap. J.*, **354**, 83.
Park, M.-G., and Ostriker, J. P. 1989, *Ap. J.*, **347**, 679.
Protheroe, R. J., and Kazanas, D. 1983, *Ap. J.*, **265**, 620.
Shakura, N. I., and Sunyaev, R. A. 1973, *Astr. Ap.*, **24**, 337.
Shapiro, S. L., Lightman, A. P., and Eardley D. M. 1976, *Ap. J.*, **204**, 187.
Shapiro, S. L., and Teukolsky, S. A. 1983, *Black Holes, White Dwarfs and Neutron Stars: The Physics of Compact Objects* (Wiley).
Sikora, M., Begelman, M. C., and Rudak, B. 1989, *Ap. J. (Letters)*, **341**, L33.
Sikora, M., Kirk, J. G., Begelman, M. C., and Schneider, P. 1987, *Ap. J. (Letters)*, **320**, L81.
Sikora, M., and Zbyszewska, M. 1985, *M. N. R. A. S.*, **212**, 553.
Stern, B., Svensson, R., and Sikora, M. 1991, in *Variability of Active Galactic Nuclei*, eds. H. R. Miller and P. J. Wiita (Cambridge University Press) p. 229.
Stern, B., and Svensson, R. 1991, this volume.
Svensson, R. 1982, *Ap. J.*, **258**, 335.
Svensson, R. 1984, *M. N. R. A. S.*, **209**, 175.
Svensson, R. 1987, *M. N. R. A. S.*, **227**, 403.
Svensson, R. 1990, in *Physical Processes in Hot Cosmic Plasmas*, eds. W. Brinkmann *et al.* (Kluwer Academic), p. 357.
Tritz, B., and Tsuruta, S. 1989, *Ap. J.*, **340**, 203.
Turner, T. J., and Pounds, K. A. 1989, *M. N. R. A. S.*, **240**, 833.
White, T. R., and Lightman, A. P. 1989, *Ap. J.*, **340**, 1024.
White, T. R., and Lightman, A. P. 1990, *Ap. J.*, **352**, 495.
Zdziarski, A. A. 1985, *Ap. J.*, **289**, 514.
Zdziarski, A. A. 1986, *Ap. J.*, **305**, 57.
Zdziarski, A. A. 1988, *M. N. R. A. S.*, **233**, 739.

Limit Cycles in Electromagnetic Cascades in Compact Objects

Boris Stern [1], Roland Svensson [2]

[1] Institute for Nuclear Research, Academy of Sciences of the USSR,
Moscow 117312, USSR
[2] Stockholm Observatory, S-133 36 Saltsjöbaden, Sweden

Abstract: Electromagnetic cascades possibly occurring near accreting compact objects have been discovered to show limit cycle behaviour. The power from accelerated protons gets converted by the cascade into soft radiation (X-rays and below) if the photon compactness is sufficiently large. Then the proton-photon system may develop limit cycles much like a prey-predator system with each component interchangebly dominating. This causes periodic large amplitude short time variability of the nonthermal luminosity from a compact object even if the acceleration or injection process is completely steady. Results both from detailed Monte Carlo simulations and from a simple phenomenological model are presented.

1. Introduction

Ultrarelativistic protons may be accelerated by shocks in accretion flows (Ellison and Eichler 1984, Blandford and Eichler 1987) onto supermassive black holes in compact objects such as active galactic nuclei (e.g. Kazanas and Ellison 1986, Zdziarski 1986). The ultra relativistic protons mainly interact with the soft radiation field in AGNs starting hadron-electromagnetic cascades (Sikora *et al.* 1987, Mannheim and Biermann 1989). In the hadron part of the cascade neutrons and pions are produced. The neutrons escape. The pions, on the other hand, decay producing ultrarelativistic γ-rays, electrons and positrons, which initiate electromagnetic cascades if the photon compactness,

$$\ell_\gamma \equiv \frac{L}{R}\frac{\sigma_T}{m_e c^3},$$

(1)

where L and R are the photon luminosity and size of the source, respectively, is large enough ($\gtrsim 50$). Then the γ-ray power is converted into soft radiation below $m_e c^2$ (e.g. Stern 1985, Zdziarski and Lightman 1985) resulting in X-ray spectra similar to those observed from AGNs (e.g. Svensson 1987, 1990; Lightman and

Zdziarski 1987; Ghisellini 1989, Done, Ghisellini and Fabian 1990; Zdziarski *et al.* 1990). The pion-decays also give rise to an appreciable neutrino luminosity.

Although the properties of electromagnetic cascades are well known, all work on the hadron part of the hadron-electromagnetic cascade has so far been based on very approximate analytical estimates of resulting particle, photon, and neutrino spectra and their respective luminosity. Also, the necessary feedback due to the conversion of γ-rays into X-rays by the electromagnetic cascade and the resulting decrease in the proton cooling time has not been accounted for. Mannheim and Biermann (1989) speculated that this feedback might naturally give rise to nonsteady behaviour of the source.

2. Monte Carlo Simulations

We consider a spherical region of size R, where confined monoenergetetic protons of Lorentz factor γ_i are uniformly injected (and possibly accelerated) with a total power L_{inj}, and where soft photons (with energies in the eV-keV-range) likewise are uniformly injected with power L_s. Fully timedependent Monte Carlo calculations were performed using large particle techniques (see Stern 1988) of the hadron-electromagnetic cascades occuring in such a region including most of the detailed physics of the most important processes. The computer code can simulate nonlinear cascades that are nonuniform in space as each large particle has its own position and direction. Our spherical region was divided into a central region with 5 surrounding shells with the injection or acceleration taking place in the central region and the innermost shell. We followed the time evolution of of the cascade and calculated the spectral evolution of all components including the escaping photons, neutrons and neutrinos.

We found that for some of the seven cases we studied the confined proton source is unstable and develops a limit cycle behavior (see Figs. 1 and 2). The photon-proton system acts like a prey-predator system with the prey and predator population interchangebly dominating. The proton energy density (the prey) builds up slowly producing γ- and X-rays (the predators). At some point the compactness is sufficient for a rapid redistribution through the electromagnetic cascade of γ-ray power into X-rays. The proton then rapidly cools (gets eaten) on the X-rays and the photon production rate drops. The photons escape and the protons can again start building up their energy density.

Figure 1 shows a case where the protons are injected at $\gamma_i = 5.4 \times 10^5$ until $t = 30R/c$ at which time the injection Lorentz factor was decreased to $\gamma_i = 1.7 \times 10^5$ keeping the total injection power constant at $\ell_{inj} = 133$ (see (1) for definition of compactness).

The upper panel in Fig. 1 shows the time evolution of the total energy contents of protons, photons and electrons/positrons in the injection region of size R. The total energy contents in the source have been made dimensionless by converting into compactness units. The 'proton energy compactness', e.g., becomes

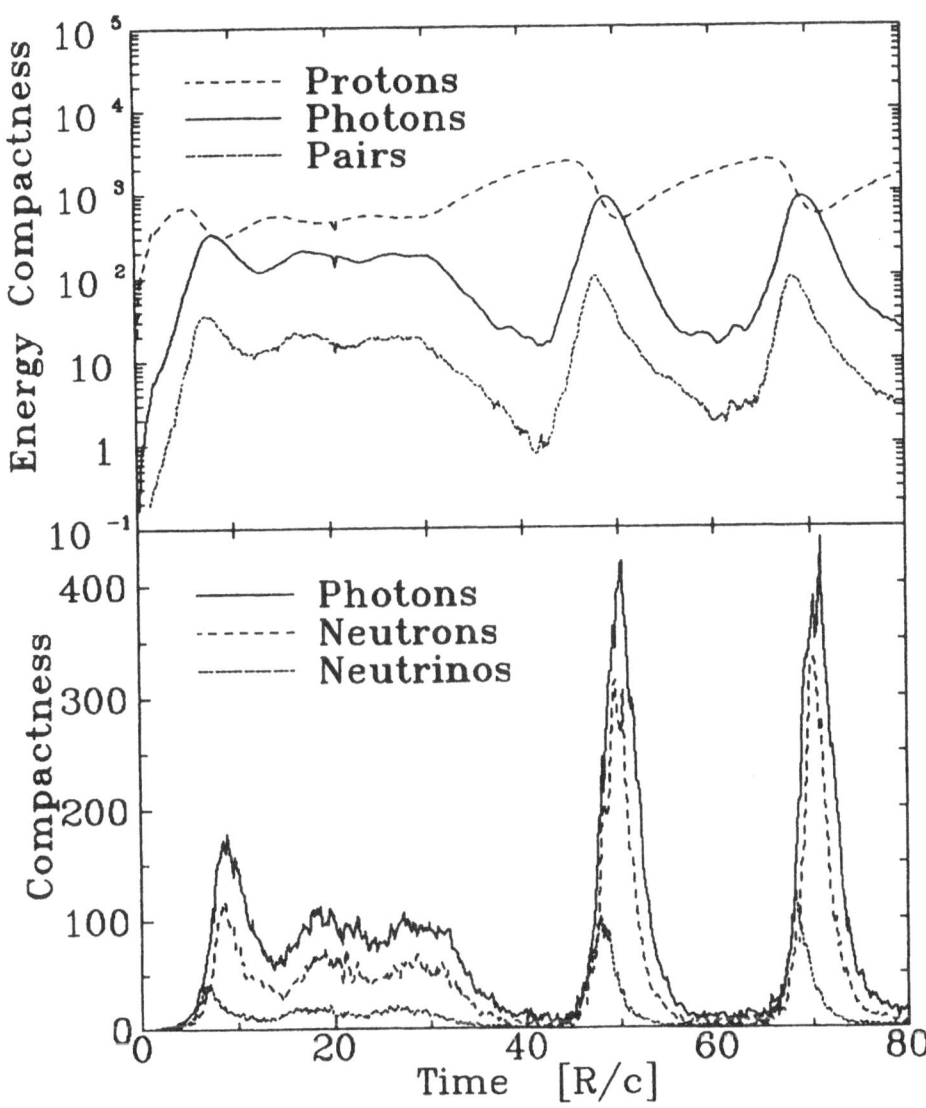

Fig. 1. Upper panel shows the time evolution of the total energy contents for the case where protons are injected at $\gamma_i = 5.4 \times 10^5$ until $t = 30R/c$ at which time the injection Lorentz factor was decreased to $\gamma_i = 1.7 \times 10^5$ keeping the injected proton power constant ($\ell_{inj}=133$). The compactness of the external soft photon injection, ℓ_s, was 1.3, and the magnetic field was set to 10^4 gauss. The lower panel shows the corresponding time evolution of the total luminosities. For the larger Lorentz factor the source relaxes to a steady state but when the Lorentz factor is decreased a limit cycle behavior develops.

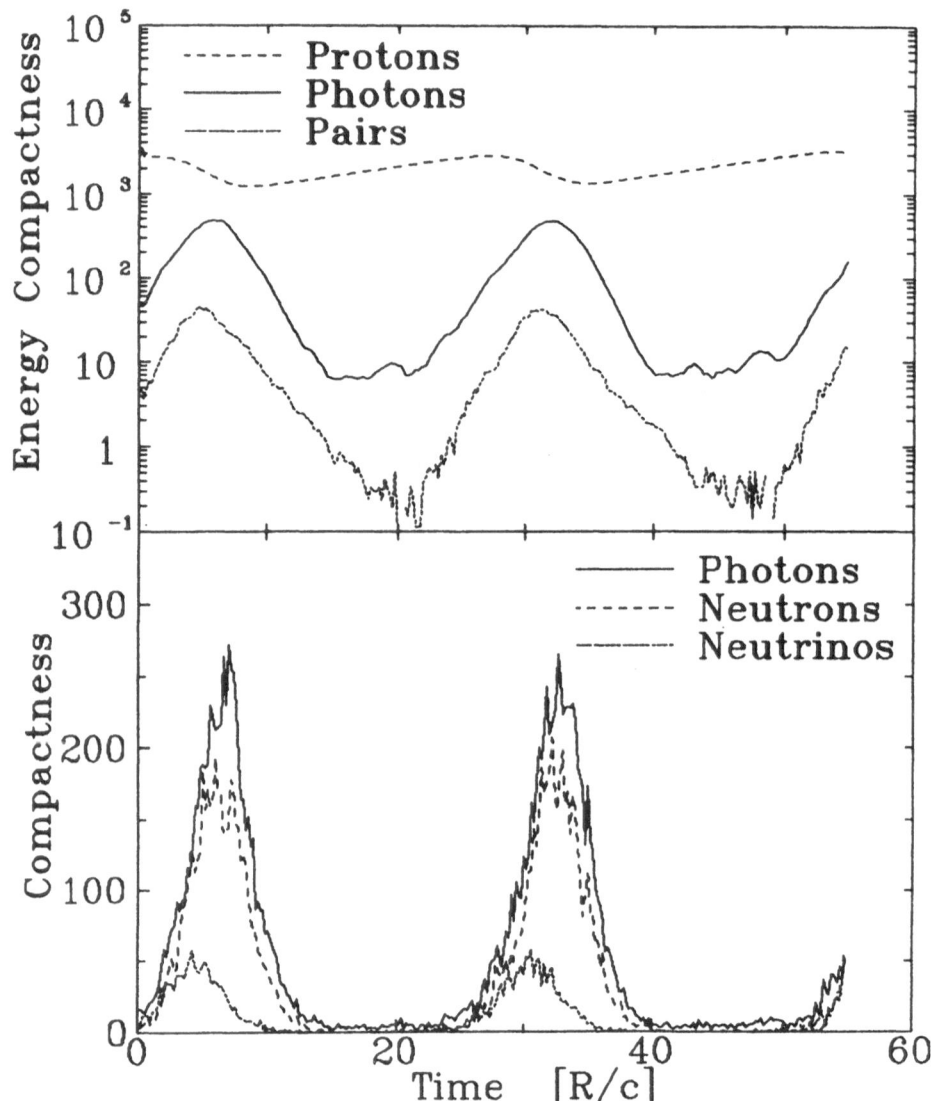

Fig. 2. Same as Fig. 1 but for the case where protons are injected at a small Lorentz factor γ_i=540 and then accelerated with an e-folding time of $0.05R/c$. The compactness of the external black body soft photon injection, ℓ_s, was 2.7 at a temperature $kT = 10^{-5}m_ec^2$.

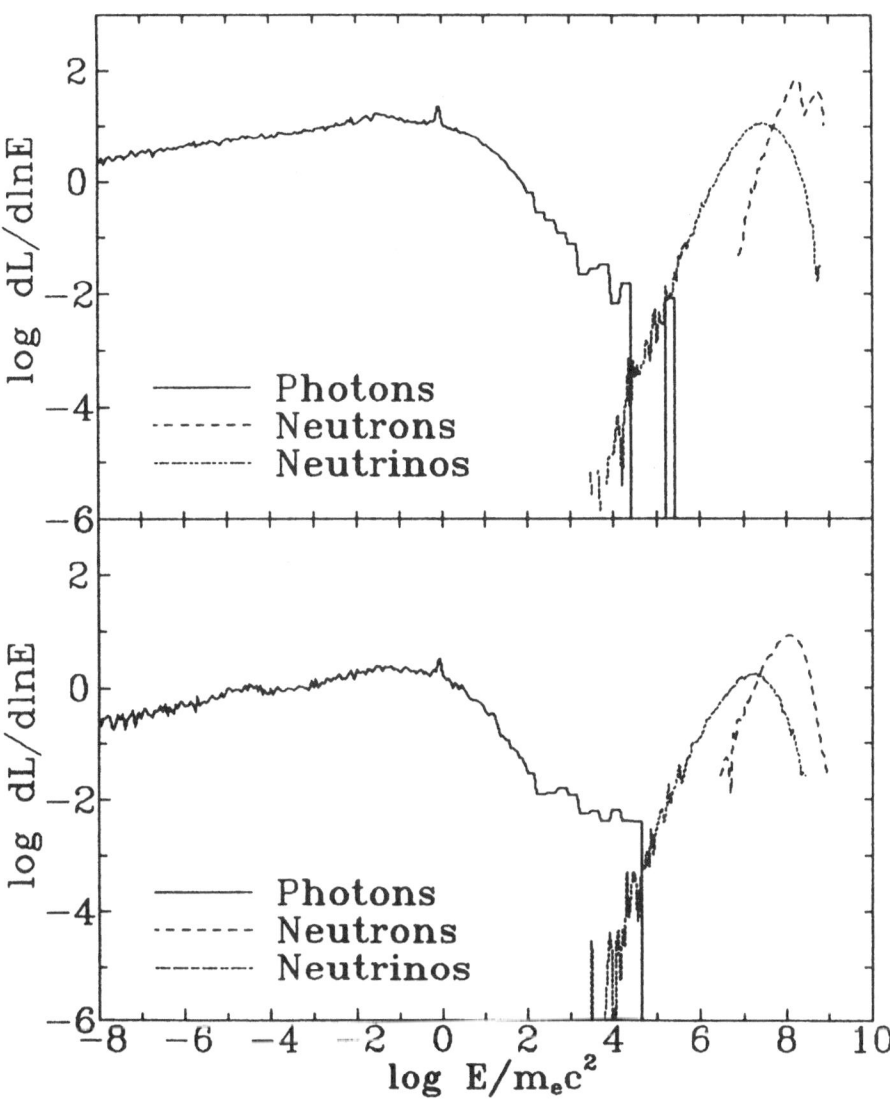

Fig. 3. Time averaged escaping photon (solid curve), neutron (dotted curve), and neutrino (dashed curve) spectra for the full runs shown in in Figs. 1 and 2. The spectra are shown as the spectral compactness per logarithmic energy interval (in arbitrary units) as a function of dimensionless energy (in units of $m_e c^2$).

$$\mathcal{E}_p = \frac{E_p c}{R^2} \frac{\sigma_T}{m_e c^3} , \qquad (2)$$

where E_p is the total proton energy content. The lower panel of Fig. 1 shows the 'bolometric light curves' for escaping photons, neutrons, and neutrinos. The system initially relaxes to a steady state through damped oscillations. After γ_i is decreased at $t = 30R/c$ a clear limit cycle behavior of the nonlinear hadron-electromagnetic cascade develops with a period of about $20R/c$.

Figure 2 displays the results of a more complicated simulation, where protons were accelerated by first order Fermi-acceleration. The protons were injected at some small Lorentz factor, $\gamma_i = 540$, and then accelerated with an energy increase $\Delta E = (\Delta t/t_{acc})\gamma_p m_p c^2$ at every time step Δt. Here, γ_p is the Lorentz factor of the proton and $t_{acc} = 0.05R/c$ is an energy-independent e-folding acceleration time. Also for this case limit cycles appear. Figure 3 shows the photon, neutron, and neutrino spectra averaged over the full runs of Figs. 1 and 2.

The typical periods, $\sim 20 - 50R/c$, we obtain for the pulsating photon source correspond to time scales of minutes to days for sources with typical sizes of 10 Schwarzschild radii around black holes with masses $10^6 - 10^9 \, M_\odot$.

Although we only peformed a limited set of timeconsuming runs it is clear that the average photon compactness must be in the range 10 - 100 for the source to show limit cycle behavior. Only in this range can the electromagnetic cascade change from not operating to becoming very efficient (in converting γ-rays into X-rays) for a small increase in compactness. We also empirically find that when the Lorentz factor of the injected protons is large enough the source becomes stable.

3. A Toy Model

In order to crudely understand the cause for the limit cycle and its dependence on various input physics, we developed a phenomenological model for hadron-induced electromagnetic cascades.

The two main components to follow are the protons (being the component receiving the initial power) and the photons (being the dominant final component). The protons are most easily described with their 'energy compactness', $\mathcal{E}_p$, defined by (2) representing the total proton energy content in the source. The photons are most conveniently described with their compactness, ℓ_γ, as given by (1). The toy model attempts to describe the time evolution of these two quantities.

The time (in units of R/c) evolution of $\mathcal{E}_p$ is given by

$$\frac{d\mathcal{E}_p}{dt} = \ell_{inj} - \frac{\mathcal{E}_p}{t_{\gamma p}} , \qquad (3)$$

where the first term on the right hand side describes injection of proton power. The second term describes the proton cooling, with

$$t_{\gamma p}^{-1} \sim 10^{-6} \gamma_i \ell_X \left(\epsilon \lesssim \frac{m_p}{m_e \gamma_i} \right) \qquad (4)$$

being the approximate proton cooling time (in units R/c) from Sikora *et al.* (1987). Here, ℓ_X is the compactness of the soft photons below the pion production threshold which are the main cooling targets for the protons. Equation (3) then takes the form

$$\frac{d\mathcal{E}_p}{dt} = \ell_{inj} - a\ell_X \mathcal{E}_p,$$ (5)

where $a \equiv 10^{-6}\gamma_i$.

For a total photon compactness, ℓ_γ greater than a critical compactness, $\ell_c \sim 50 - 100$, there is a total redistribution of the γ-ray ($\epsilon > 1$) luminosity into X-ray ($\epsilon < 1$) luminosity by the electromagnetic cascade (Svensson 1987, Lightman and Zdziarski 1987). Here, then ℓ_X is a fraction $c \sim 0.1$ of ℓ_γ. For small $\ell_\gamma \ll \ell_c$ there is no redistribution and ℓ_X makes up a much smaller fraction (by a factor $x \sim 0.1$) of ℓ_γ, i.e. $\ell_X = xc\ell_\gamma$. A simple prescription for $\ell_X(\ell_\gamma)$ having these limits is

$$\ell_X(\ell_\gamma) = c\ell_\gamma \left[\frac{x + (\ell_\gamma/\ell_c)^d}{1 + (\ell_\gamma/\ell_c)^d} \right],$$ (6)

where d is a measure a how rapid the transition between the two limiting cases occur.

The time evolution of ℓ_γ is given by

$$\frac{d\ell_\gamma}{dt} = fa\ell_X(\ell_\gamma)\mathcal{E}_p - \ell_\gamma,$$ (7)

where the first term describes the photon luminosity gain as being a fraction $f \sim 0.5 - 0.8$ of the proton losses (second term in (5)). The second term in (7) describes photon escape, with the escape time in units of R/c being unity (photon diffusion is neglected in this version of the toy model).

Equations (5)-(7) now completely describes the evolution of the system. There is a total of seven parameters. Four of them, $x \sim 0.1$, $c \sim 0.1$, $d \sim 2-4$, and $\ell_c \sim 100$ is set by the theory of electromagnetic cascades (Svensson 1987, Lightman and Zdziarski 1987). In the hadron part of the cascade $f \sim 0.5 - 0.8$. The only free parameters are $a \propto \gamma_i$ and ℓ_{inj}.

The equilibrium solutions are given by

$$\ell_\gamma = f\ell_{inj},$$ (8)

and

$$\mathcal{E}_p = \frac{\ell_{inj}}{a\ell_X(\ell_\gamma)} = \frac{1+y}{acf(x+y)},$$ (9)

where

$$y \equiv \left(f\frac{\ell_{inj}}{\ell_c} \right)^d.$$ (10)

A linear stability analysis of (5)-(7) shows that the system is stable if

$$\kappa \equiv \frac{acf(x+y)^2\ell_{inj}}{d(1-x)y} > 1,$$ (11)

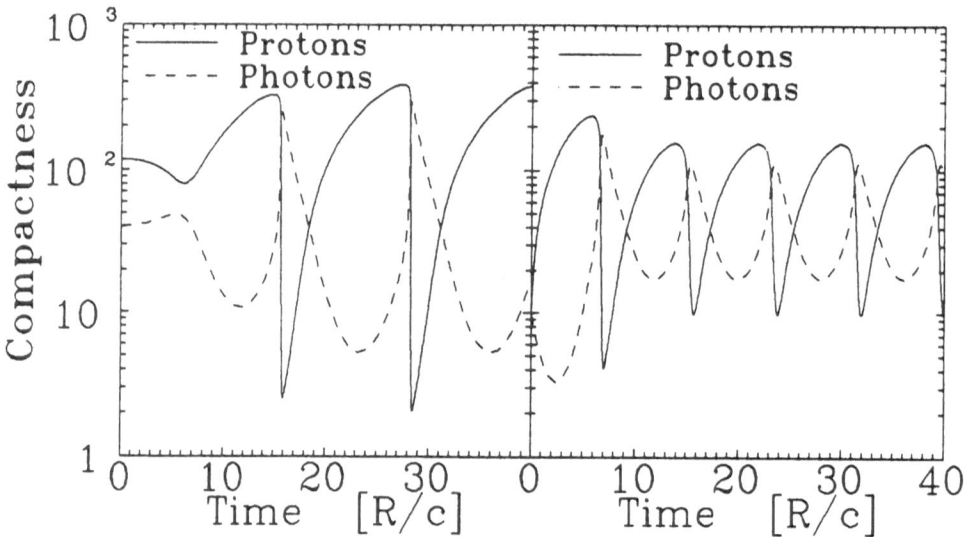

Fig. 4. Evolution $\mathcal{E}_p$ and ℓ_γ of toy model with $a=1$, $\ell_{inj}=50$, $f=0.8$, $c=0.1$, $d=3$, and $\ell_c=100$. In the left panel, $x=0.05$ and the instability parameter, $\kappa=0.29$, while in the right panel, $x=0.1$ and $\kappa=0.62$.

Figure 4 show the evolution of $\mathcal{E}_p$ and ℓ_γ for the above set of parameters (with $d=3$), $a=1$, $\ell_{inj}=50$ and for two choices of x, 0.05 and 0.1. The stability parameter, κ, for the two cases are 0.29 and 0.62, respectively. The system is thus unstable and settles into a limit cycle. Both periods and amplitudes are sensitive to the details of the electromagnetic cascade such as a factor of two change in x. Figure 5 show the evolution towards a limit cycle in configuration space, $\mathcal{E}_p - \ell_\gamma$, for the same cases as in Fig. 3. The dashed curves in Fig. 5 show where the time derivatives of $\mathcal{E}_p$ and ℓ_γ are zero. The crossing of the dashed curves is the equilibrium solution given by (8) and (9). In the left panel of Figs. 4 and 5 the system is started very near the equilibrium point, but it rapidly evolves towards the limit cycle.

The stability parameter, $\kappa \propto a \propto \gamma_i$. Increasing the injection Lorentz factor may therefore stabilize the system. Figure 6 shows the cases where a was increased from 1.5 to 1.625 (with the remaining parameters being the same as above and $x=0.1$) making κ increase from 0.93 across the stability threshold to 1.01. In the latter case the system evolves very slowly towards its stable equilibrium state.

From the stability parameter (11) it is clear that a system without redistribution of γ-rays into X-rays (which can be modelled with $x=1$) is always stable. The redistribution by the electromagnetic cascade is therefore the essential ingredient in obtaining instability and limit cycle behaviour.

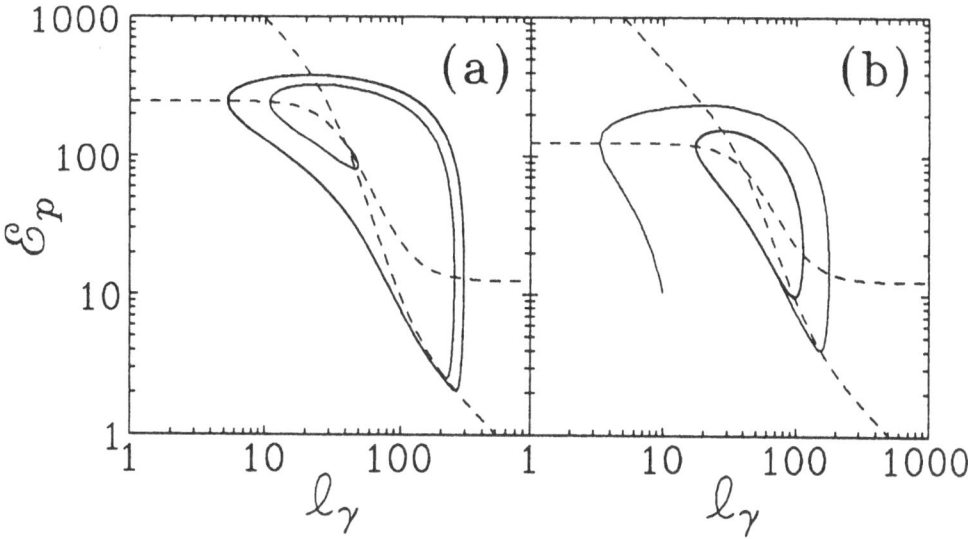

Fig. 5. Evolution in configuration space for the cases in Fig. 4. Dashed curves are curves of zero time derivatives for $\mathcal{E}_p$ and ℓ_γ.

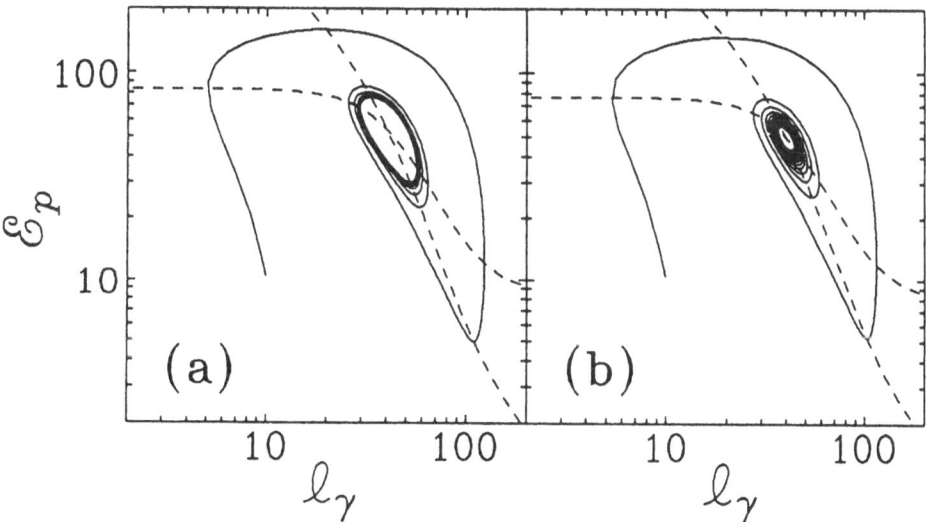

Fig. 6. Evolution in configuration space for two cases with parameters as in Fig. 4, but with $x=0.1$ and $a=1.5$ (and thus $\kappa=0.93$) in Fig. 6a, and $a=1.625$ (and thus a stable $\kappa=1.01$) in Fig.6b.

4. Summary

We have shown that a *radiation mechanism* can exhibit time variability although the external physical conditions were assumed to be steady. The physical conditions are not expected to remain constant and a more realistic model will show more complex time variability.

Blandford, R. D., and Eichler, D. 1987, *Phys. Rep.*, **154**, 1.
Done, C., Ghisellini, G., and Fabian, A. C. 1990, *M. N. R. A. S.*, **245**, 1.
Ellison, D. C. and Eichler, D. 1984, *Ap. J.*, **286**, 691.
Ghisellini, G. 1989, *M. N. R. A. S.*, **238**, 449.
Kazanas, D., and Ellison, D. C. 1986, *Ap. J.*, **304**, 178.
Lightman, A. P., and Zdziarski, A. A. 1987, *Ap. J.*, **319**, 643.
Mannheim, K., and Biermann, P. I. 1989, *Astr. Ap.*, **221**, 211.
Sikora, M., Kirk, J. G., Begelman, M. C., and Schneider, P. 1987, *Ap. J. (Letters)*, **320**, L81.
Stern, B.E. 1985, *Sov. Astr.*, **29**, 306.
Stern, B.E. 1988, NORDITA preprint 88/51.
Svensson, R. 1987, *M. N. R. A. S.*, **227**, 403.
Svensson, R. 1990, in *Physical Processes in Hot Cosmic Plasmas*, eds. W. Brinkmann, A. C. Fabian, and F. Giovannelli (Dordrecht: Kluwer Academic), p. 357.
Zdziarski, A. A. 1986, *Ap. J.*, **305**, 45.
Zdziarski, A. A., and Lightman, A. P. 1985, *Ap. J. (Letters)*, **294**, L79.
Zdziarski, A. A., Ghisellini, G., George, I. M., Svensson, R., Fabian, A. C., and Done, C. 1990, *Ap. J. (Letters)*, **363**, L1.

Nuclear Interactions Near Accreting Black Holes

V. I. Dokuchaev

Institute of Nuclear Research, USSR Academy of Science
60th Anniversary of October Revolution Prospect 7a
117312 Moscow, USSR

The possibility of nuclear interactions near a black hole has been widely discussed since the work of Dahlbacka, Chapline and Weaver (1974). They noticed that those interactions are likely to proceed at the near-virial temperatures of $T_p \sim 10^{12}$ K, which can be attained by accreting matter. Particularly high proton temperatures are attained in the case of fast rotating black hole. Particles of gas near the inner edge of a thick accretion disk move essentially along the nonequatorial geodesics. The instant planes of these geodesics precess around the rotation axis of the black hole (Wilkins 1972). As a result, the nonequatorial orbits are entangled by this precession and the absolute values of plasma particle velocities in the meridianal direction achieve $v \leq (2/3)^{1/2} c$ (Dokuchaev 1986). This is enough for pion generation in pp inelastic collisions. The subsequent decay of these pions will produce neutrinos and γ-quanta with the average energy of $E \sim 100$ MeV. The maximum energy of the nuclear interaction products escaping to infinity is determined by the Penrose process and could approach 1 GeV. The thickness of the disk near its inner edge depends on the permissible range of polar angle θ. For the stable nonequatorial orbits, $\cos \theta \geq \cos \theta_0 = (2 \cdot 3^{1/2} - 3)^{1/2}$, $\theta_0 \simeq 47°$. The plasma column density in such a disk is $\sim (m_p/\sigma_T \epsilon)(L/L_E) \sim 25(L/L_E)$ g cm^{-2}, where the accretion efficiency $\epsilon = L/\dot{M}c^2 \sim 0.1$. It appears that pp collisions could produce a noticeable fraction of the total luminosity, L, in high energy γ-rays and neutrinos if the accretion process is sufficiently close to the Eddington limit, L_E.

The exact values of high energy γ-ray and neutrino luminosities crucially depend on the accretion pattern near the black hole. The simplest case is a "slow accretion" (Berezinsky and Dokuchaev 1990), where the energy loss time of protons due to Coulomb interactions with electrons, τ_{pe}, is much longer than the characteristic infall time, $\tau_{\inf} = r/v(r)$. In this regime, which is realized if $\dot{M} \lesssim 10^{-3} \dot{M}_E$, where $\dot{M}_E = L_E/\epsilon c^2$, nearly all protons move along the regular trajectories and frictionally spiral down through a thermalized electron gas with $kT_e \sim m_e c^2$ near the black hole. In this fully computable toy model of globally spherically symmetric slow accretion onto a nonrotating black hole, the high energy γ-ray luminosity is $L_\gamma \simeq 5 \times 10^{-2} L$. It is interestig to note that the efficiency of spherical slow accretion, ϵ, is as high as in a geometrically thin disk because nearly all spiralling down protons draw their own imaginary accretion disks. Only about every 50th of the slowly accreting protons gives rise to a γ-quantum after an inelastic pp collision, independent of the total accretion rate $\dot{M}$.

References

Berezinsky, V. S., and Dokuchaev, V. I. 1990, *Ap. J.*, **361**, 492.
Dahlbacka, J. H., Chapline, G. F., and Weaver, T. A. 1974, *Nature*, **250**, 37.
Dokuchaev, V. I. 1986, *Sov. Astron. Lett.*, **15**, 322.
Wilkins, D. C. 1972, *Phys. Rev. D*, **5**, 814.

A New Approach to Hot Accretion Disks

Gunnlaugur Björnsson [1] [3], Roland Svensson [2]

[1]Nordita, Blegdamsvej 17, DK-2100 Copenhagen Ø, Denmark
[2]Stockholm Observatory, S-133 36 Saltsjöbaden, Sweden
[3]Supported in part by the Icelandic Science Foundation

Abstract: We present a new method for investigating hot two-temperature geometrically and effectively optically thin accretion disks including effects at mildly relativistic temperatures such as electron-positron pair production. We solve the disk structure equations analytically and show that the many studies of hot plasma clouds existing in the literature can be used directly to analyse the detailed disk structure. A number of new results were discovered using our approach. In particular we find equilibrium solutions at all radii and for all accretion rates, and we have discovered a critical value of the viscosity parameter, above which physically acceptable pair dominated disk solutions exist.

1 Introduction

We consider the two-temperature optically and geometrically thin accretion disk of Shapiro, Lightman & Eardley (1976), but include electron-positron pair production in the manner of Svensson (1982, 1984). Due to the non-linearity of the pair balance equation when photon-photon interactions dominate the pair production, this problem is usually tackled numerically (Kusunose & Takahara 1988, 1989, Tritz & Tsuruta 1989, White & Lightman 1989).

In this paper we present a new method for investigating the structure of these hot, two-temperature accretion disks, including electron-positron pairs. We build on, and directly utilize, the many studies of hot plasma clouds that exist in the literature. In fact, we show that there is a simple correspondence between the pair equilibria as studied in the works of e.g. Svensson (1982, 1984), Lightman (1982), Zdziarski (1985), Sikora & Zbyszewska (1986), Björnsson & Svensson (1991a), and the solutions obtained in the numerical approaches referenced above. However, our method is a considerable improvement over a direct numerical solution of the problem as it allows a visualization of all disk solutions at once in a single figure, and provides a number of new insights into the properties of optically thin accretion disks.

2 Method and Results

The physical properties of hot spherical plasma clouds are mainly determined by *two* parameters. These are the 'proton optical depth' (the Thomson scattering depth of the ionized electrons), τ_p, and the *global* compactness parameter, $\ell \equiv (L/H)(\sigma_T/m_e c^3)$, where L and H denote the luminosity and size of the cloud, respectively.

By imagining the disk to consist of small plasma clouds at each radius (with the size of each cloud being the scale height, H, of the disk), we can directly use the results of the hot plasma cloud studies within the accretion disk models. This is accomplished by observing that $\tau_p \propto \Sigma$, the disk surface density, and that the compactness, ℓ, at a given radius is determined by the luminosity emitted locally by a surface of size H^2.

Using these parameters as the independent variables of the problem, the disk equations can be solved analytically (Björnsson 1990; Björnsson & Svensson 1991b),

$$\dot{m}\left(\frac{r}{r_S}\right)^{-3/2}\phi(r) = \frac{\sqrt{2}}{\pi}\frac{m_e}{m_p}\frac{\ell}{(P/\rho c^2)^{1/2}}, \qquad (1)$$

$$\alpha = \frac{1}{2\pi}\frac{m_e}{m_p}\frac{\ell}{\tau_p(P/\rho c^2)^{3/2}}. \qquad (2)$$

Here, $\dot{m} \equiv \dot{M}/\dot{M}_{Edd}$, $r_S = 2GM/c^2$ and $\phi(r) = 1 - \sqrt{3r_S/r}$ describes the stress free boundary condition at the disks inner edge. These two relations hold for all α-disks independent of the (chosen) microphysics. When the microphysics has been specified, the pressure to rest mass energy density ratio, $P/\rho c^2$, can easily be calculated for a given equilibrium state of the plasma. Utilizing the above expressions considerably simplifies the numerical approach to the problem as only the equations describing the microphysics need to be solved numerically. The expressions provide a direct mapping of a given microphysical equilibrium state into an equilibrium disk state at some specific $\dot{m}(r/r_S)^{-3/2}\phi(r)$ (i.e. radius) and α. The solutions of the microphysics equations in terms of τ_p and ℓ are well studied in the hot plasma cloud studies in the literature.

Our main results are (Björnsson 1990; Björnsson & Svensson, 1991b):

i) The solutions from the hot plasma cloud studies can easily be used to create a surface describing *all* disk solutions. It allows the solutions to be analyzed at all radii and for all accretion rates at once. The topography of the surface directly shows how the disk structure changes as the viscosity parameter, α, changes. The surface for Comptonized bremsstrahlung disks is shown in Fig. 1. Using the results of earlier hot plasma cloud studies, the regions on the surface where the various processes dominate can be straight-forwardly identified.

ii) The radius and the accretion rate, $\dot{m}$, always occur together, as in the combination $\dot{m}(r/r_S)^{-3/2}\phi(r)$, (see (1) and the axis label in Fig. 1). Hence, for a fixed α there is a generic solution profile that is obtained from the surface in Fig. 1. This generic profile can be used to calculate the full disk solution for any

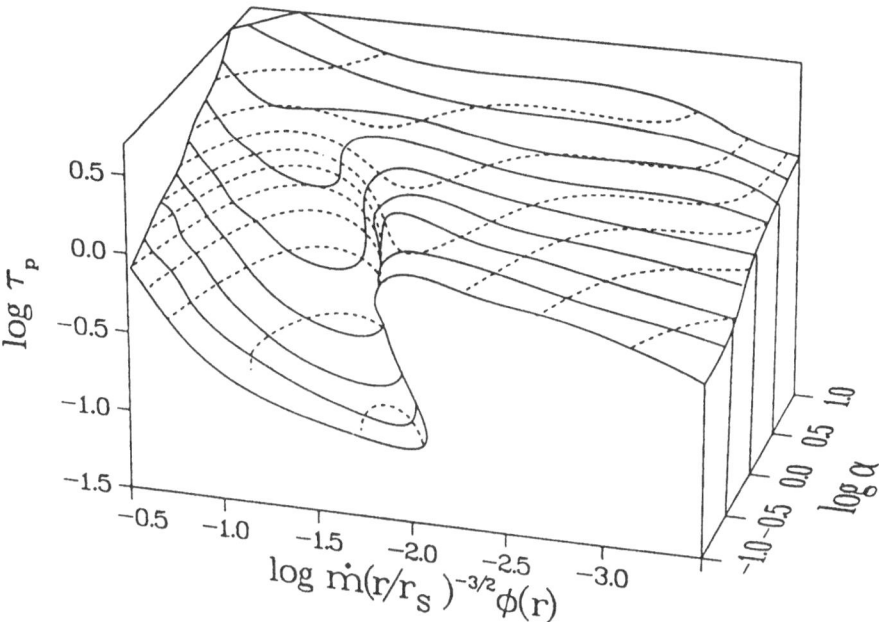

Fig. 1. The surface of all equilibrium states of hot plasma clouds as well as of all accretion disk solutions. The dashed curves show, for constant τ_p, the solutions of the microphysical equations, while the solid curves show the generic accretion disk profiles each corresponding to a different value of α. The 'valley' to the left is pair dominated, while the 'slope' to the right is essentially pair free. The stress free boundary condition is accounted for by $\phi(r) = 1 - \sqrt{3r_S/r}$, where $r_S = 2GM/c^2$.

$\dot{m}$, using a simple coordinate transformation. An example of a full disk solution obtained in this manner is given in Fig. 2.

iii) In the case of Comptonized bremsstrahlung disks, there is a fold in the surface of equilibrium states for $\alpha < \alpha_{cr} \sim 1$ causing the existence of *two* radius dependent critical accretion rates, a lower, $\dot{m}_L$, and an upper, $\dot{m}_U$ (of order unity). For $\dot{m} < \dot{m}_L$ only *one* physical solution exists at all radii. It is essentially pair free. For accretion rates inbetween the two critical rates, regions in the disk have *three* equilibrium solutions (see Fig. 2), two of which are unphysical as the proton temperature is much higher than the virial temperature. For $\dot{m} > \dot{m}_U$ electron-positron pairs cause infinite radial gradients to appear and a physically acceptable solution exists only where there are few pairs. Figure 2 shows the disk surface density profiles ($\Sigma \propto \tau_p$) for three different accretion rates. Profiles with similar topological behavior are obtained for other physical quantities.

iv) For $\alpha > \alpha_{cr}$ the fold in the surface in Fig. 1 disappears and only one solution exists at all radii and for all accretion rates. Detailed calculations of the profile for $\alpha = 2.0$ show that for moderately high accretion rates ($\dot{m} \sim$ a few), the

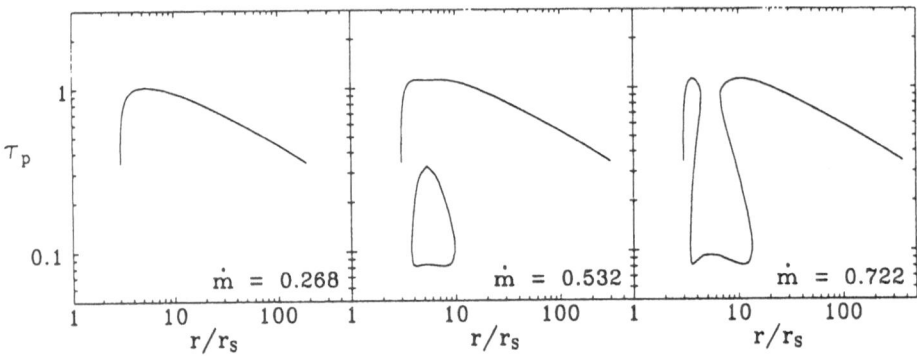

Fig. 2. The disk profiles for the proton optical depth, $\tau_p \propto \Sigma$, for the case of $\alpha = 0.1$. In the leftmost panel $\dot{m} < \dot{m}_L$. In the middle panel $\dot{m}_L < \dot{m} < \dot{m}_U$, while in the rightmost panel $\dot{m} > \dot{m}_U$. Note the approximate symmetry around $r/r_S \sim 5$.

solution in the innermost part of the disk is pair dominated (with $n_+/n_p \sim 2$) and is consistent with the assumptions of the geometrically thin accretion disk.

References

Björnsson, G., 1990, Ph.D. Thesis, University of Illinois.
Björnsson, G. & Svensson, R., 1991a, *Mon. Not. R. astr. Soc.*, in press.
Björnsson, G. & Svensson, R., 1991b, in preparation.
Kusunose, M. & Takahara, F., 1988, *Publ. Astr. Soc. Japan*, **40**, 435.
Kusunose, M. & Takahara, F., 1989, *Publ. Astr. Soc. Japan*, **41**, 263.
Lightman, A. P., 1982 *Ap. J.*, **253**, 842.
Shapiro, S. L., Lightman, A. P. & Eardley, D. M., 1976, *Ap. J.*, **204**, 187.
Sikora, M., & Zbyszewska, M., 1986, *Acta Astr.*, **36**, 255.
Svensson, R., 1982, *Ap. J.*, **258**, 335.
Svensson, R., 1984, *Mon. Not. R. astr. Soc.*, **209**, 175.
Tritz, B. & Tsuruta, S., 1989, *Ap. J.*, **340**, 203.
White, T. R. & Lightman, A. P., 1989, *Ap. J.*, **340**, 1024.
Zdziarski, A. A., 1985, *Ap. J*, **289**, 514.

X-Ray Continuum Emission Caused by
Relativistic Protons

Karl Mannheim

Max-Planck-Institut für Radioastronomie
Auf dem Hügel 69, D-5300 Bonn 1
Federal Republic of Germany

Abstract: In this contribution the case is made that relativistic protons as predicted by diffusive shock acceleration can lead to X-ray emission exceeding synchrotron-self-Compton emission by electrons accelerated by the same process. Photomeson cooling of protons on ambient soft synchrotron photons of the primary electrons initiates a three-generation cascade of secondary particles terminating at X- to soft γ-ray energies. Pure nonthermal sources like hot spots in Fanarov-Riley type 2 (FR II) galaxies or similar objects are proposed as optimum candidates to confirm predicted convex spectra with an effective spectral index $\alpha_x = 0.7^{+0.2}_{-0.1}$ in the ROSAT-band and a luminosity ratio $L_{\mathrm{keV}}/L_{<1\,\mathrm{eV}} \approx 1\%$.

1. Introduction

The workshop addressed the problem of relativistic hadrons in astrophysics. Whereas hadronic cosmic rays in the Galaxy are known to exert a pressure comparable to the magnetic field pressure and high energy γ-ray point sources have been detected, there is not a single piece of empirical evidence confirming the existence of relativistic hadrons in extragalactic objects conclusively. Although sources of *soft* γ-rays associated with compact objects have been observed outside the Galaxy, they do not neccessarily relate to fast ions. Nevertheless, both facts, compact relativistic proton sources presumably feeding an extended component, are sometimes used to propose cosmic rays in other galaxies where we can measure neither relativistic ions *in situ* nor use diffuse γ-ray emission as a tracer. Yet science is different: We do not draw conclusion from observations, but from theories explaining observations. So what does theory tell us about ions?

In a painstaking two-decade effort the theory of diffusive shock acceleration has been developed explaining nonthermal polarized continuum emission from a shocked tenuous magnetized plasma flow (e.g., Drury 1983, Webb, Drury and Biermann 1984). "Hot spots", i.e. high surface-brightness regions in the extended radio lobes associated to FR II galaxies, represent a unique laboratory for nonthermal processes, in particular for particle acceleration and losses (see also Krülls, this volume). Roughly speaking, whenever electrons or positrons are accelerated by resonant scattering off magnetic irregularities as inferred from synchrotron radiation, so are ions. However, the acceleration process depends on details like the type of scattering centers responsible for pre-acceleration and acceleration, the structure of the shock wave, the level of turbulence, the streaming pattern of the plasma flow, etc.. Differing in mass hadrons "see" different waves and structures compared to leptons. Hence in general it is not possible to predict ion spectra and flux densities by simple analogy to leptons, even if parameters of the acceleration theory can be tuned by virtue of the knowledge of synchrotron emission from electrons observed at radio to optical frequencies. An additional constraining set of assumptions must be included into the description to couple the problem of accelerated leptons and hadrons.

This has been done by Biermann and Strittmatter (1987) introducing the assumption of saturated Kolmógorov turbulence excited by upstream protons (presumably the most abundant hadrons). As an immediate consequence

(i) the ubiquitous near-infrared to optical cutoff frequency around 3×10^{14} Hz in hot spots and BL Lacs can be explained independent of the magnetic field strength in agreement with observations.

Saturated turbulence means equal energy densities in ordered and fluctuating magnetic fields $u_{\delta B} = u_B$ which implies that the energy density of wave exciting protons must satisfy $u_p \gtrsim u_B$. On the other hand, to confine relativistic particles magnetically $u_p \lesssim u_B$ is required. Hence also $u_p \approx u_B$ must be assumed. Diffusive shock acceleration theory does not neccessarily contradict saturated turbulence, since the diffusion coefficient can be rather robust to the level of turbulence.

Another far reaching consequence is the prediction of a flat spectrum of extremely energetic protons. In fact, energies up to roughly 10^{20} eV can be reached in hot spots. Remarkably, this prediction – seemingly implausible at first sight – is still consistent with two further important facts:

(ii) Cosmic rays of this energy are seen by atmospheric shower detectors (Yakutsk, Haverah Park and Fly's Eye). The radius of gyration of a proton at 10^{20} eV in the interstellar magnetic field is about 30 kpc inconsistent with a galactic origin of protons.

(iii) Hot spots in FR II galaxies typically have diameters of a few kpc slightly exceeding the radius of gyration of a proton at a few 10^{20} eV moving through a magnetic field of about 30 nT. This magnetic field strength was derived by two independent methods (low frequency break, minimum energy with $\eta = u_p/u_e = 1 \cdots 100$ for high loss hot spots; see Meisenheimer et al., 1989).

Hence the scenario of dynamical equilibrium between protons and magnetic field ($\eta > 1$ so that $u_B \approx u_p$) and turbulence (saturation, $u_{\delta B} = u_B \approx u_p$) is consistent with observations and raises the question: Can we see the theoretically expected protons storing large amounts of energy?

In fact, this contribution is devoted to an answer of this question and it turns out that the answer is – yes. Let us see, how this conclusion emerges. Firstly, it is well-known that a relativistic proton population immersed in a soft photon field (viz. synchrotron photons from the accelerated electrons and virtual photons of the magnetic field) suffers losses by synchrotron emission, Bethe-Heitler pair production and photomeson production (Biermann and Strittmatter 1987, Sikora et al. 1987, Mannheim and Biermann 1989). Secondly, if the target photons obey an inverse power law (as they do), the pair creation opacity of γ-rays resulting from the secondary particles increases with energy. Since the photon density in hot spots is much lower than in compact nuclear sources, we know that the opacity exceeds one only at very high energies. However, the emissivity of secondary particles produced by a flat proton distribution peaks at the highest energies. Thus, with a maximum proton energy of about 10^{20} eV most of the power in secondaries is nevertheless reprocessed by further scattering, i.e. pair creation followed by synchrotron emission. The second generation photons may still be energetic enough to be further reprocessed, hence an electromagnetic cascade shifts the secondary particle energy distribution rapidly towards lower energies (compare with Compton cascades, Svensson 1987).

In the following chapters we will give the kinetic equations governing the coupled transport of electrons/positrons and photons, derive their solution in terms of an integral equation and discuss the results for high loss hot spots.

2. The Coupled Transport Equations

A linear decomposition of the coupled transport problem of waves, protons and electrons injected from the thermal plasma and secondary particles is attempted. The heuristic ansatz assumes protons to adjust dynamical equilibrium with self-excited isotropic Alfvén waves with an intensity spectrum of Kolmógorov type. Shock acceleration in the strong nonrelativistic limit transfers kinetic energy from the jet flow to a nonthermal flat proton distribution. Electrons scatter from the same wave spectrum to also develop a flat power law distribution breaking by one power due to strong synchrotron losses. Cutoff energies $h\nu_c$ and $m_p c^2 \gamma_{p,max}$ can now be calculated explicitly (see Biermann and Strittmatter 1987, equations (21),(22) and (29)) yielding[1] for a shock speed of $c/\sqrt{3}$

$$\nu_c = 3 \times 10^{14} \text{ Hz} \tag{1}$$

and

$$\gamma_{p,max} = 1.7 \times 10^{10} B^{-0.5} \left[1 + 240a\right]^{-0.5} \tag{2}$$

where

$$a = u_\gamma / u_B \tag{3}$$

The photon to magnetic field energy ratio also determines the ratio of synchrotron-self-Compton (SSC) to synchrotron luminosity. For hot spots $a \ll 1$ – in marked

---

[1] CGS units throughout

contrast to compact nuclear emission regions. The electron spectral break will be absent for protons, since the fractional energy loss in the coupled scenario is always much smaller for protons, viz.

$$\epsilon = \frac{u_{sec}/u_p}{u_{syn}/u_e} = \frac{1}{\eta}\frac{L_p}{L_e} = const \ll 1 \tag{4}$$

Now, whereas the electron synchrotron radiation dominates the low energy regime $\nu \lesssim 10^{14}$ Hz, the proton secondaries cause radiation with maximum power at $\nu \gg 10^{14}$ Hz. Hence we can introduce the *target* field number distribution[2] $m_t(x)$ representing synchrotron radiation from an electron distribution $n_t(\gamma)$ by

$$m_t(x) = m_{0t}\, x^{-2} \quad \text{for} \quad x_b \le x \le x_c \tag{5}$$

The target distributions will then serve for calculating scattering probabilities of high energy electrons/positrons (treated in the Born approximation) with distribution $n(\gamma)$ and photons with distribution $m(x)$ in a linear fashion. The photomeson emissivities can be calculated integrating over the proton distribution, viz. for $\alpha_t = 1$ we have

$$q_\pi(\gamma_\pi) = C_\pi(u_{\gamma t}, u_{et}, \eta)\, \gamma_\pi^{-1} \quad \text{for} \quad 10^8 \le \gamma_\pi \le 2\gamma_{p,max} \tag{6}$$

where

$$C_\pi(u_{\gamma t}, u_{et}, \eta) = \frac{4.8 \times 10^{-3}\langle\sigma_{p\gamma}\rangle u_{\gamma t} u_{et}\eta}{m_p m_e c^3 \ln[x_c/x_b]\ln[\gamma_{p,max}/\gamma_0]} \tag{7}$$

see Mannheim and Biermann (1989) for details. Here we used the isobar resonance $\Delta(1234)$ dominating the scattering probability of protons cooling in inverse power law target fields (5), therefore $\langle\sigma_{\gamma p}\rangle \approx 200\,\mu$b. Further inserting decay distributions for $\pi^0 \to 2\gamma$ and $\pi^\pm \to e^\pm 3\nu$ (Ramaty and Lingenfelter 1966; Stecker 1971) we obtain

$$q_{e\pm}(\gamma) = C_{e\pm}\gamma^{-1} \quad \text{for} \quad 1.4 \times 10^{10} \le \gamma \le 140\gamma_{p,max} \tag{8}$$

and

$$q_\gamma(x) = C_\gamma x^{-1} \quad \text{for} \quad 2.6 \times 10^{10} \le x \le 260\gamma_{p,max} \tag{9}$$

where simply

$$C_{e\pm} = C_{\pi\pm} = \frac{1}{2}C_\pi \quad \text{and} \quad C_\gamma = 2C_{\pi^0} = C_\pi \tag{10}$$

The emissivities $q_{e\pm}$ and q_γ represent injection terms based upon diffusive shock acceleration theory to be inserted into the general Boltzmann equations for secondary particles. Various assumptions enter into the description:

- The target and injection distributions are simplified assuming a sharp cutoff and neglecting the part below the given minimum energies which are actually break energies. The former does not affect the power law continuum which we are interested in and the latter is justified by the fact that the maximum power lies at highest energies.

---

[2] energy density $u = m_e c^2 \int xm(x)dx$; $x = \frac{h\nu}{m_e c^2}$

- Spherical symmetry, isotropy and homogenity can be assumed, because high loss hot spots occupy regions in space with almost equal length in perpendicular directions, we do not consider highly relativistic shocks and a leaky-box description is applied (Jones 1970). Proper escape times assure the breaks in the spectra due to pair creation and synchrotron losses.
- To neglect Fermi acceleration we restrict the validity of the results to $x \gg x_c$ and $\gamma \gg \gamma_c$, respectively, whereby the corresponding terms act only as a perturbation. By construction of the model x_c is determined by equal acceleration gains and synchrotron and Compton losses.
- Proton synchrotron radiation and Bethe-Heitler pair production are not considered, because the corresponding luminosities peak at very high γ-ray energies, though not high enough for reprocessing to be important.

Hence we obtain the coupled kinetic equations

$$\frac{\partial n(\gamma,t)}{\partial t} - C_{syn,1}\frac{\partial \gamma^2 n(\gamma,t)}{\partial \gamma} + \frac{n(\gamma,t)}{T_{esc}^{(e)}} = q_e(\gamma,t) + 4Q_{\gamma\gamma}[m](2\gamma,t) \qquad (11)$$

and

$$\frac{\partial m(x,t)}{\partial t} + \frac{m(x,t)}{T_{esc}^{(\gamma)}} = q_\gamma(x,t) + Q_{syn}[n](x,t) - Q_{\gamma\gamma}[m](x,t) \qquad (12)$$

where

$$Q_{\gamma\gamma}[m](x,t) = \frac{c}{R}\tau(x)\,m(x,t) \qquad (13)$$

and

$$Q_{syn}[n](x,t) = C_{syn,2}x^{-1/2}\,n\left(\gamma(x)\right) \qquad (14)$$

with

$$\gamma(x) = 10^7 B_\perp^{-\frac{1}{2}} x^{\frac{1}{2}} \qquad (15)$$

and the pair creation opacity

$$\tau(x) = C_{\gamma\gamma}x \qquad (16a)$$

where

$$C_{\gamma\gamma} = 4 \times 10^{-31}\frac{L_t}{R\ln x_c/x_b} \qquad (16b)$$

where R denotes the source radius and L_t the luminosity of photons with $\nu \leq \nu_c$. The coefficients are $C_{syn,1} = 1.3 \times 10^{-9}B^2$ and $C_{syn,2} = 9 \times 10^{11}B^{0.5}$.

Equation (11) gives the changes of the $e^\pm$-energy distribution due to losses by particle escape and synchrotron radiation and gains through secondary particle injection and photon pair creation. Correspondingly, (12) accounts for changes in the high energy photon distribution due to escape and pair creation losses and synchrotron and injection gains.

Classical synchrotron formula are still applicable, because in hot spots the characteristic photon energy

$$x_{syn} = 3.4 \times 10^{-14}B_\perp\gamma^2 \qquad (17)$$

satisfies the condition $x_{syn} \ll \gamma$ for $\gamma \ll 10^{17}$. Assuming emission of synchrotron photons only at the maximum of the radiated distribution we obtain the synchrotron emissivity

$$q_{syn} = Q_{syn}[n](x) = C_{syn,2}x^{-0.5}n[\gamma(x)] \tag{18}$$

3. The Solution

Clearly, the stationary solution of equation (12) is given by

$$m(x) = \frac{R}{c}\frac{Q_{syn}[n](x) + q_\gamma(x)}{\tau(x) + \lambda(x)} \quad \text{for} \quad x \gg x_c \tag{19}$$

where λ denotes the photon escape probability due to pair creation

$$\lambda = \tau[\exp[\tau] - 1]^{-1} \tag{20}$$

whence we see that $[\tau + \lambda]^{-1}$ is just $[1 - \exp[-\tau]]/\tau$ or with sufficient accuracy $[1 + \tau]^{-1}$.

Equation (19) has a very simple meaning: The stationary $e^\pm$-distribution gives rise to synchrotron radiation and photoproduction leads to additional γ-radiation with both components breaking by one power above x^* due to pair creation.

The method of solving (19) is to first derive the homogenous solution of (11) and thereof construct the general solution of (11) including the injection terms by the convolution theorem. Power law injection (8) yields

$$n(\gamma) = \frac{1}{C_{syn,1}\gamma^2}\exp\left[-\frac{\gamma_b}{\gamma}\right]\left\{C_{e\pm}\Gamma\left[0, -\frac{\gamma_b}{\gamma_{max}}, -Min\left[\frac{\gamma_b}{\gamma_{min}}, \frac{\gamma_b}{\gamma}\right]\right]\right.$$
$$\left. + 8C_{\gamma\gamma}\frac{c}{R}\int_\gamma^\infty dp\, p\exp\left[\frac{\gamma_b}{p}\right]m(2p)\right\} \tag{21}$$

where

$$\gamma_b = \frac{1}{C_{syn,1}T_{esc}^{(e)}} \ll \gamma_c \tag{22}$$

The symbol Γ denotes the generalized incomplete Gamma function

$$\Gamma[a, b, c] = \int_b^c dp\, p^{a-1}\exp[-p]$$

Omitting pair production, viz. $C_{\gamma\gamma} = 0$, equation (21) is the stationary solution of electron transport with synchrotron and escape losses together with power law injection (Fig. 1).

Finally, inserting (21) into (19) we obtain an integral equation for the stationary photon distribution

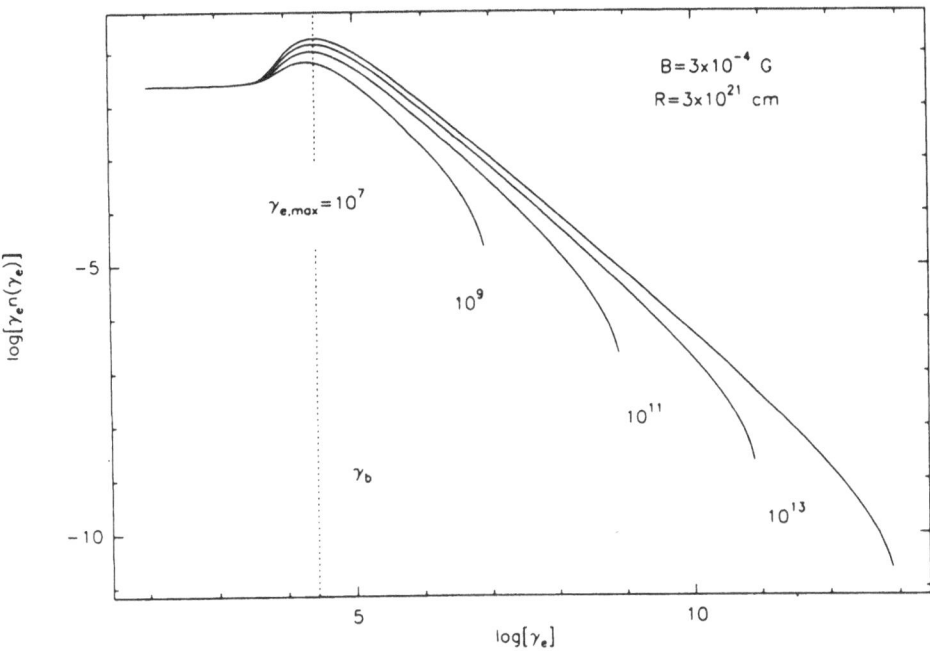

Fig. 1. The stationary $e^{\pm}$ distribution with synchrotron losses and power law injection from the $\pi^{\pm}$-decay.

$$m(x) = g(x) + \int_x^\infty dp\, k(x,p)\, m\left[2\gamma(p)\right] \tag{22}$$

The inhomogenous part $g(x)$ and the kernel $k(x,p)$ are given by

$$g(x) = \frac{R}{c}\frac{1}{\tau(x) + \lambda(x)} \left\{ C_\gamma x^{-1} H\left[x_{\min}^{\pi^0} x_{\max}^{\pi^0}\right] + C_{e\pm} C_{\mathrm{syn},1}^{-1} C_{\mathrm{syn},2} x^{-\frac{1}{4}} \gamma(x)^{-2} \right.$$

$$\left. \times \exp\left[-\frac{\gamma_b}{\gamma(x)}\right] \Gamma\left[0, -\frac{\gamma_b}{\gamma_{\max}}, -Min\left[\frac{\gamma_b}{\gamma_{\min}}, \frac{\gamma_b}{\gamma(x)}\right]\right] H\left[x_{\min}^{\pi^\pm} x_{\max}^{\pi^\pm}\right] \right\} \tag{23}$$

and

$$k(x,p) = \frac{4}{\tau(x) + \lambda(x)} C_{\gamma\gamma} C_{\mathrm{syn},1}^{-1} C_{\mathrm{syn},2} x^{-\frac{1}{2}} p^{-1} \left[\frac{\gamma(p)}{\gamma(x)}\right]^2 \exp\left[\frac{\gamma_b}{\gamma(p)} - \frac{\gamma_b}{\gamma(x)}\right] \tag{24}$$

Equation (22) can be solved iteratively by

$$m(x) = \sum_{n=0}^{\infty} m_n(x) \tag{25}$$

so that

$$m_n(x) = \int_x^\infty dp\, k(x,p)\, m_{n-1}\left[2\gamma(p)\right] \qquad (26)$$

with

$$m_0(x) = g(x) \qquad (27)$$

Convergence $\sum_n^N m_n \to m\ (N \to \infty)$ is achieved rapidly, since already the third generation m_3 turns over below the energy $x^* \approx 10^8$ where the source becomes opaque so that no further cascading occurs. E.g., the maximum energies of the first three π^0-cascade generations of a typical high loss hot spot ($L_t = 10^{44}$ erg s^{-1}, $R = $ kpc and $B = 3 \times 10^{-4}$ G) are 8×10^{13}, 5×10^9 and $19 \ll 10^8$.

In the next section we will demonstrate the solution for the prominent hot spot 3C273 A.

4. Application

Figure 2 depicts the luminosity per logarithmic bandwidth diagram for hot spot A of the bright quasar 3C273. The hot spot is very similar to high loss hot spots in FR II galaxies with respect to spectrum, polarization, size, magnetic field strength, particle density and luminosity. The separation of more than 15 arcsec from the nucleus makes X-ray observations possible as demonstrated by Harris and Stern (1986) using Einstein HRI data. The predicted spectrum is based upon physical parameters as derived by multifrequency observations (see Meisenheimer et al. 1989) where η was chosen such that the downstream estimate from the low frequency break equals the minimum energy field strength, i.e. $B_{me} = B_+$.

The linear synchrotron cascade of shock accelerated particles (LSCSP) obviously causes a convex X- to soft γ-ray spectrum turning over at a few MeV. The luminosity in the ROSAT band is roughly 1% of the radio to optical luminosity exceeding the synchrotron-self-Compton emission (SSC) by a factor of 5. Also the keV spectral index $\alpha_{LSCSP} = 0.63$ differs significantly from the corresponding SSC-index $\alpha_{SSC} = 1$. There is some uncertainty – apart from the truth of the theoretical premises – due to the crude treatment of the source geometry. For instance, these uncertainties allowing for some 10% deviations from the calculated luminosities could result in almost equal keV-luminosities for the two physical processes. Hence we are left with the spectral index. The combined spectral index of SSC and LSCSP in Fig. 2 is $\alpha_x = 0.7$ where X denotes the energy range $0.1 - 2.5$ keV. The 10% uncertainty in L_{SSC}/L_{LSCSP} then yields the result $\alpha_x = 0.7^{+0.2}_{-0.1}$. The behaviour over the entire X-ray band should reveal the origin of the X-rays unambiguously, since the pure SSC spectrum tends to $\alpha = 1$ towards higher energies, whereas the combined spectrum remains close to 0.7.

Bremsstrahlung from the hot spot itself has a comparatively low luminosity ($n_e = 10^{-3}$ cm^{-3} and $T = 10^{10}$ K lead to $L_{brems} \approx 10^{38}$ erg s^{-1}), whereas the surrounding medium certainly radiates considerable bremsstrahlung. The similar hot spots of Cyg A have not been seen by the Einstein HRI, because even if they emitted 10^{41-42} erg s^{-1}, the count rate in the 12 arcsec beam is dominated

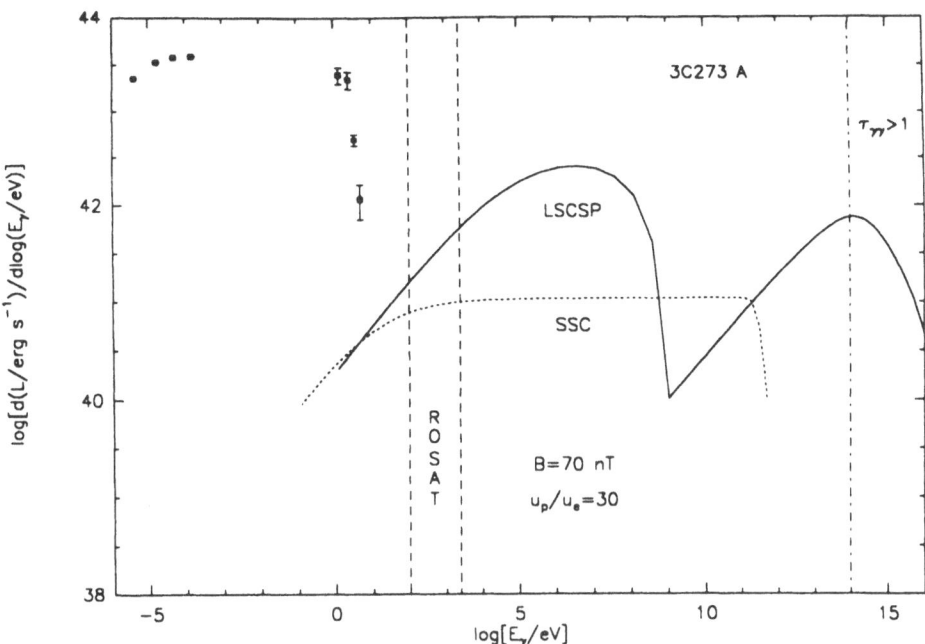

Fig. 2. The luminosity per logarithmic bandwidth spectrum of the high loss hot spot 3C273 A. Data are from Meisenheimer et al. (1989). LSCSP denotes the predicted continuum radiation caused by a flat relativistic proton distribution, SSC the competing synchrotron-self-Compton radiation.

by thermal bremsstrahlung from cooling intracluster gas (Arnaud 1984). Galaxies not located in a cluster are therefore much better suited for a hot spot detection. So high resolution ROSAT-observations are required to distinctively measure X-rays from hot spots.

It is unclear, whether the X-ray feature of roughly 8×10^{43} erg s^{-1} at 15.8 arcsec from the nucleus of 3C273 claimed by Harris and Stern (1986) is true. If it were, both the luminosity and morphology do not completely agree with the simple picture proposed. The same is true for M87 A and the questionable detection of Pic A west where the X-ray luminosities are much higher than expected from SSC or LSCSP. Detailed spectra are needed to settle these questions.

The notable γ-ray luminosity is still below the GRO detection limits. The contribution to the γ-ray background could be considerable, however. The predicted LSCSP emission is consistent with the MeV-bump of the extragalactic background emission. Another bump close to 100 TeV can be predicted, if the number of comparable hot spots is sufficient. However, statistical conclusions should not be anticipated, since so far we have considered sources similar to high loss hot spots of FR II galaxies only. Whereas low loss hot spots are not important due to the paucity of soft target photons, more compact objects like VLBI knots or blazars self-produce most target photons for pairproduction, so that solution (22) assuming a linear cascade is invalid. Yet SCSP could still be important, if $\eta \gg 1$.

5. Discussion

Cascade emission caused by secondary particles produced by highly relativistic protons has been described in the framework of diffusive shock acceleration theory. The cascade emission is the hadronic counterpart to the synchrotron-self-Compton emission from primary electrons.

The linear treatment restricts the applicability of the solution to either large scale luminous objects like the hot spots 3C273 A, Pic A west and 3C123 east or fainter, but smaller objects like the M87 knot A. It is shown that the predicted convex X- to γ-ray continuum exceeds SSC-emission at ROSAT energies. Also contributions to the γ-background at a few MeV and close to 100 TeV have been identified. The confirmation of an X-ray continuum with $\alpha_x = 0.7^{+0.2}_{-0.1}$ and $L_{\text{keV}} \approx 10^{-2} L_{<\text{eV}}$ could demonstrate the existence of relativistic protons in extragalactic jets, whereas a negative result would pose a severe problem with the assumption of Kolmógorov turbulence and saturation.

The Einstein data of M87 A, Pic A west and 3C273 A , if confirmed by ROSAT, suggest a much higher X-ray luminosity than expected for SSC and LSCSP. In principle these data can be fitted by cascade emission increasing the cosmic ray ratio η, by geometrical modifications or assuming a strongly beamed radiation field, so that the photon density increases. The former procedure, however, leads to very large minimum energy magnetic field strengths which seem inconsistent with the energy content of hot spots and contradict the values inferred from low frequency breaks.

References

Arnaud, K.A., Fabian, A.C., Eales, S.A., Jones, C., Forman, W. 1984, *Mon. Not. R. astr. Soc.*, **211**, 981

Biermann, P.L., Strittmatter, P.A. 1987, *Ap. J.*, **322**, 643

Drury, L.O'C. 1983, *Rep. Prog. Phys.*, *Vol.*46, 973

Harris, D.E., Stern, C.P. 1987, *Ap. J.*, **313**, 136

Jones, F.C. 1970, *Phys. Rev. D*, *Vol.*2, No. 12, 2787

Mannheim, K., Biermann, P.L. 1989, *Astr. Ap*, **221**, 211

Meisenheimer, K., Röser, H.-J., Hiltner, P.R., Yates, M.G., Chini, R., Perley, R.A. 1989, *Astr. Ap.*, **219**,63

Ramaty, R., Lingenfelter, R.E. 1966, *J. Geo. Res.*, *Vol.* 71, No. 15, 3687

Sikora, M., Kirk, J.G., Begelman, M.C., Schneider, P. 1987, *Ap. J.*, **320**, L81

Stecker, F.W. 1971, *Cosmic Gamma Rays*, Mono Book, Baltimore

Svensson, R. 1987, *Mon. Not. R. astr. Soc.*, **227**, 403

Webb, G.M., Drury, L.O'C., Biermann, P.L. 1984, *Astr. Ap*, **137**, 185

Pair Cascade Model Constraints on Energetic Particle Emission from Active Galactic Nuclei

Matthew G. Baring

1 Max-Planck-Institut für Astrophysik, Karl Schwarzschild Strasse 1, D-8046 Garching bei München, Fed. Rep. of Germany.

Abstract: The injection of extremely relativistic protons in the compact central region of an AGN creates electrons and neutrons by interaction with the ambient radiation and particles. The resulting electrons have very high Lorentz factors and participate in a Klein-Nishina pair cascade with UV photons. These high γ electrons cause a flattening of the gamma-ray spectra obtained from first-order pair cascade models, which is contrary to observations. This fact is used to place upper limits on the high γ electron injection luminosity in AGN central regions, and consequently deduce upper limits for luminosities of the relativistic neutrons that are produced and emitted from these compact regions.

1 Introduction

The idea of extremely energetic particles being emitted from the compact central region of an active galactic nucleus (AGN) has rapidly gained popularity in the last few years. Kazanas and Ellison (1986), Zdziarski (1986) and Sikora et al. (1987) proposed that shocks present in the central region could accelerate protons efficiently to Lorentz factors as large as 10^8 and be the power source for the AGN emission. The protons scatter off lower-energy (thermal) protons or the ambient radiation of the region, triggering a chain of electromagnetic and hadronic reactions with electrons, photons, pions, muons, neutrons and neutrinos as end products. The pions and muons are short-lived, while the electrons, photons and neutrons are the most interesting products. The electrons and photons become involved in pair cascades that have been modelled by numerous authors in the last five years (e.g. see Svensson, 1987; Lightman and Zdziarski, 1987). The product electrons, with their large energies, can have a profound effect on the resulting AGN gamma-ray spectrum (Baring, 1989). The neutrons are also of great interest to theorists. Being neutral, they are not bound by the electromagnetic fields and can flow freely

---

1 Present address: Department of Physics, North Carolina State University, Raleigh, NC 27695, U.S.A.

from the central region, transporting large amounts of energy to much large radii from the central black hole. This exciting possibility has spawned a number of papers addressing aspects of neutron production and transport from AGN central regions (Kirk and Mastichiadis, 1989; Mastichiadis and Protheroe, 1990; Sikora, Begelman and Rudak, 1989; Begelman, Rudak and Sikora, 1990).

The effect of the product electron injection on the pair cascade development in the central region is inextricably tied to the proton injection (which is chosen to suit individual models), to which the neutron production is also correlated. The interplay of the different species is quite complicated, particularly since the neutrons can participate in similar reactions to the protons. If the electrons injected into the compact central region interact with ambient UV photons (i.e. considering sources with low I.R. and optical fluxes there), then the ensuing cascade very probably results in a flattening in the gamma-ray spectrum (above about 1 MeV), an effect that is not observed in source spectra (Perotti, et al. 1981a, 1981b; Baity et al. 1981). This flattening is caused by a build-up of particles due to a lower cooling rate in the Klein-Nishina limit of Compton scattering (Baring, 1989; 1991). The lack of observational evidence (which in fact indicates a steepening) for this flattening suggests that the high energy electron component must not be responsible for the determination of the AGN spectrum via a pair cascade. The spectrum is best modelled with an electron injection of around $\gamma_e \sim 100$ (Done, et al. 1990). In this paper, first-order pair cascade theory is used to place upper limits on high energy electron injection luminosities (competing with a lower energy injection) resulting from energetic protons, and therefore also on the product relativistic neutron emission luminosities.

2 Proton Injection in AGNs and its Products

In the compact central region (about 30 Schwarzschild radii in size) protons are assumed to be injected with a power-law relativistic distribution $Q_p(\gamma_p) \propto \gamma_p^{-p}$ for $1 \leq \gamma_p \leq \gamma_{\text{max}}$, possibly by the first-oder Fermi shock acceleration process (e.g. see Sikora, et al. 1987, hereafter SKBS87). The following basic properties of the proton and ensuing electron injections are discussed in SKBS87, but are only summarized here. The shock acceleration process gives an injection index of $p < 2$ and a maximum proton energy of $\gamma_{\text{max}} \sim 10^8$, where the process becomes dominated by the rate of cooling of the protons in collisions with soft photons. Proton cooling by collisions with cold protons, considered by Kazanas and Ellision (1986) and Zdziarski (1986), only dominates cooling by γp collisions when $\gamma_p \lesssim 10^5$. Fig. 1 of SKBS87 shows that the acceleration timescale is much shorter than the various proton cooling times when $\gamma_p \ll 10^8$, so that generally protons escape from the shock acceleration region before cooling, thereby retaining the γ_p^{-p} distribution.

Once accelerated, the protons cool producing pions, gamma-rays, neutrons and electrons via a number of different processes. In γp collisions, which dominate for $\gamma_p \gtrsim 10^5$, there is a direct pair production channel, for which the pairs have an energy corresponding to $\gamma_e \sim \gamma_p$ and the threshold for this channel is at photon energies $2m_e c^2/\gamma_p$ (see SKBS87). The cross-section for $\gamma p \to p e^+ e^-$ is of the order of $0.003\sigma_T$ (Begelman, Rudak and Sikora, 1990; hereafter BRS90), where σ_T is the Thomson cross-section, and is a slowly increasing function of energy. Since the produced pair energies are about m_e/m_p times the proton energy, $\dot{\gamma}_p \propto \gamma_p$ for this direct pair production. When the protons produce pions via photomesonic interactions, these decay into photons (from π^0 channels) and pairs (from $\pi^\pm \to \mu^\pm \to e^\pm$), simultaneously producing neutrons and neutrinos. The pairs take a fraction of about m_π/m_p of the proton's energy, but the reaction rate is about two orders of magnitude smaller than for the direct pair production channel (SKBS87). Again $\dot{\gamma}_p \propto \gamma_p$, so that in total, the cooling of the protons results in a distribution $n_p(\gamma_p) \propto \gamma_p Q_p/\dot{\gamma}_p \propto \gamma_p^{-p}$. Since the pairs and the neutrons take approximately constant fractions of the protons energy, they too are "injected" with γ^{-p} distributions, with the total numbers of each being very similar.

In this paper, the protons will be assumed to interact predominantly with thermal ultra-violet photons, and that the infra-red and optical components of the AGN spectrum in the central region are negligible. This is a matter of convenience, following many pair cascade studies (e.g. Lightman and Zdziarksi, 1987), and is quite different from the model assumptions of SKBS87 and BRS90.

3 Pair Cascade Limits on High-Energy Particles

The essential problem with the presence of high-energy electrons in large numbers in the AGN emission region is that through a pair cascade they create a flattening of the gamma-ray spectrum, which is not observed. The physics behind this flattening is discussed fully in Baring (1989) and Baring (1991), a brief summary of which is presented here. When electrons with $\gamma_e > 1/\varepsilon_s$ collide with soft UV photons of energy $\varepsilon_s m_e c^2$ (with $\varepsilon_s \approx 10^{-4}$), the Compton scattering process is discrete, in the so-called Klein-Nishina regime, where most of the electron's energy is transferred to the photon. These photons can create pairs in collisions with the UV photons and the process repeated to create many generations of a Klein-Nishina pair cascade. The cooling rate of the electrons by Compton scattering is independent of γ in the Klein-Nishina (K-N) limit while it is proportional to γ^2 in the limit of Thomson scattering ($\gamma_e \varepsilon_s \ll 1$, e.g. see Fig. 1 of Baring, 1989). Since the electron cooling distribution has a form $n_e(\gamma_e) \propto \gamma_e^{1-\Gamma}/\dot{\gamma}_e$ for an electron injection rate $Q_e(\gamma_e) \propto \gamma_e^{-\Gamma}$ and a cooling rate $\dot{\gamma}_e$, then the electron distribution flattens by an index of two when γ_e becomes larger than $1/\varepsilon_s$ (e.g. Baring, 1989). This "overpopulation" at high energies competes with the reduced cooling rate in the Klein-Nishina limit to yield an inverse Compton scattering spectrum as shown in Fig. 2 of Baring (1989), which features a bump, or flattening, at gamma-ray energies.

The appearance of this bump, which is not tempered by the inclusion of pair production (see Fig. 1 below), is an unavoidable consequence of high-energy electron injection in the AGN central region, and is contrary to observations (see section 1). The injection of high energy electrons must therefore be small compared with a much lower-energy component (with $\gamma_e \lesssim 10^3$ for example), such as that used by Done et al. (1990) to model the AGN X-ray and gamma-ray continuum. The pair cascade spectra in Figure 1 were obtained by solving (see Baring, 1989) the steady-state kinetic equations for the electron and photon distributions, assuming monoenergetic UV photon and power-law electron injections and that the electron injection compactness was low enough for multiple inverse Compton scatterings of photons to be unimportant (i.e. assuming cascades that are first-order in Compton scattering).

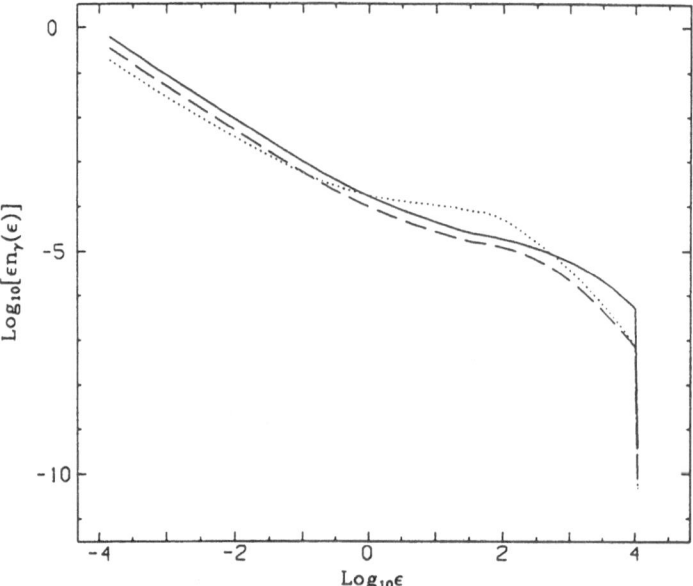

Fig. 1. The pair cascade spectra (taken from Baring, 1989) resulting from a steady-state cooling power-law electron distribution of injection index $\Gamma = 2$, revealing a Compton "bump" at high energies ($\epsilon \gtrsim 1$) when Klein-Nishina effects are important. The solid, dashed and dotted curves correspond to injection compactnesses $l_e = 4$, 40 and 400, respectively, for a monoenergtic UV soft photon injection at $\epsilon_s = 10^{-4}$ of compactness $l_s = 5l_e$ (see Lightman and Zdziarski 1987, for definitions of l_e and l_s). The curves are suitably normalized for illustrative clarity. Above $\epsilon = 10^4$, the cascade saturates because of copious pair production of gamma-rays colliding with UV photons.

Suppose that a $\gamma_e \lesssim 10^3$ injection component caused by some unspecified acceleration mechanism and a high-energy electron (pair) injection from proton collisions with protons and photons occur simultaneously. Then the nature of the flattening illustrated in Figure 1 can be used to place a limit on the high-energy electron injection luminosity. Analytic estimates are best obtained by neglecting pair production and using the pure Compton cascade results of Baring (1991; see his Fig. 2): for $\Gamma \approx 2$, the purely inverse Compton spectra have the steady-state asymptotic forms

$$n_\gamma(\varepsilon) \sim \begin{cases} 6 \times 10^{-4} \, l_e \, \varepsilon_s^{(\Gamma-2)/2} \varepsilon^{-(\Gamma+2)/2} \,, & \varepsilon \ll 1/\varepsilon_s \,; \\ 1.5 \, l_e \, \varepsilon^{-\Gamma} \,, & \varepsilon \gg 1/\varepsilon_s \,. \end{cases} \qquad (1)$$

Here l_e is the dimensionless electron injection luminosity (or compactness) as defined in Lightman and Zdziarski (1987): $l_e \sim 10^3$ corresponds roughly to an Eddington luminosity. The AGN spectrum results from a low-energy injection with luminosity l_e (determining the shape below 100 MeV) and the high-energy electron injection component with luminosity l_{eh} (providing a high energy tail to the spectrum).

How large l_{eh} can be without violating the spectral observations is now easily estimated, to order of magnitude. The spectrum at $\varepsilon = 1$ (which is the energy of UV photons inverse-Compton scattered by electrons with $\gamma_e \sim 100$) is obtained from equation (1) from the low-energy injection component, and is $n_\gamma(1) \sim 0.0006 l_e \varepsilon_s^{(\Gamma-2)/2}$. The spectrum at $\varepsilon = 1/\varepsilon_s \approx 10^4$ (i.e. in the Klein-Nishina regime) is obtained from the high-energy component of injection, and is $n_\gamma(10^4) \sim 1.5 l_{eh} \varepsilon_s^{\Gamma}$. The average spectral (power) index spanned by this gamma-ray range is then

$$\alpha_\gamma \approx 0.25 \log_{10}\left(\frac{4 l_e}{l_{eh}}\right) - \left(1 + \frac{\Gamma}{8} \log_{10} \varepsilon_s\right) \,. \qquad (2)$$

Quite clearly, if observational spectra require that $\alpha_\gamma \gtrsim 1.5$, when $\Gamma \approx 2$ it follows that $l_{eh}/l_e \lesssim 10^{-6}$, a severe constraint on the high-energy injection component. Refinements of this estimate using computed spectra rather than the asymptotic limits of equation (1) may lead to an increase in the bound to l_{eh}/l_e by an order of magnitude. The inclusion of pair production can also relax this constraint by about an order of magnitude, so that $l_{eh}/l_e \lesssim 10^{-4}$ is an approximate upper bound.

Restricting l_{eh} obviously limits the initial proton injection luminosity. Since these protons are confined to the central region, and therefore are not observable at larger radii, the upper limit to their injection luminosity is not of great interest here. However, the produced neutron luminosity is. It is simply coupled to l_{eh} since (see section 2) an injected proton colliding with UV photons transfers about half of its energy to product neutrons and about m_e/m_p to electrons when via direct pair production (the pion channels have greater inelasticity but correspondingly smaller reaction rate). Therefore $l_n \sim 10^2 l_{eh} \lesssim 10^{-2} l_e$, and it follows that less than about 10^{-2} times the AGN luminosity can be expelled from the central region in the form of neutrons. In order to avoid such a tight constraint, a high-energy proton injection is required in a region where there is a paucity of soft photons and that is also different from the region of pair cascade development. This may be quite difficult to achieve. Perhaps higher-order pair cascade models with the inclusion of multiple Compton scatterings and Comptonization of the gamma-ray continuum might reduce the severity of this constraint. Note also that neutron escape from the central region can be inhibited by its own participation in photomeson interactions, further reducing the emitted luminosity of neutrons

(Kirk and Mastichiadis, 1989; BRS90). Finally note that if infra-red and optical photons are present in the compact central region, the injected high-energy electrons initiate a pair cascade with only Thomson scattering occuring. Therefore, no spectral flattening results. However, it might still be difficult to produce the pair production gamma-ray turnover at 1 MeV that is demanded by observations.

4 Conclusion

In this paper it is observed that significant high-energy proton injection in the compact central region of an AGN would produce so many extremely relativistic electrons that a flattening in the soft gamma-ray portion (1 MeV to 1000 MeV) of the (first-order Compton) pair cascade AGN spectrum would occur. Since no such flattening is observed, it is deduced that the fractional luminosity of $\gamma > 10^3$ electrons injected into the emission region may be lower than 10^{-4} and that neutrons produced by the protons, which are subsequently emitted from the central region of the AGN, probably carry no more than 10^{-2} of the AGN luminosity. More refined modelling may alter these constraints somewhat.

References

Baity, W. A., et al.: 1981, *Ap. J.* **244**, 429

Baring, M. G.: 1989, in *Proc. 23rd ESLAB Symposium on Two Topics in X-ray Astronomy* held in Bologna, Italy, Sept. 1989 (ESA, Noordwijk, SP-296), Vol. 2, p. 891

Baring, M. G.: 1991, *M.N.R.A.S.*, to be submitted

Begelman, M. C., Rudak, B., and Sikora, M.: 1990, *Ap. J.* **362**, 38 (BRS90)

Done, C., Ghisellini, G., and Fabian, A. C.: 1990, *M.N.R.A.S.* **245**, 1

Kazanas, D., and Ellison, D. C.: 1986, *Ap. J.* **304**, 178

Kirk, J. G. and Mastichiadis, A.: 1989, *Astr. Ap.* **211**, 75

Lightman, A. P., and Zdziarski, A. A.: 1987, *Ap. J.* **319**, 643

Mastichiadis, A. and Protheroe, R. J.: 1990, preprint.

Perotti, F., et al.: 1981a, *Ap. J. (Lett.)* **247**, L63

Perotti, F., et al.: 1981b, *Nature* **292**, 133

Sikora, M., Begelman, M. C. and Rudak, B.: 1989, *Ap. J. (Lett.)* **341**, L33

Sikora, M., Kirk, J. G., Begelman, M. C., and Schneider, P.: 1987, *Ap. J. (Lett.)* **320**, L81 (SKBS87)

Svensson, R.: 1987, *M.N.R.A.S.* **227**, 403

Zdziarski, A. A.: 1986, *Ap. J.* **305**, 45

Jets and Hotspots in
Extragalactic Radio Sources

U. Achatz [1], H. Lesch [2], R. Schlickeiser [1]

[1] Max-Planck-Institut für Radioastronomie, Auf dem Hügel 69, D-5300
Bonn 1, F.R.G.

[2] Landessternwarte Königsstuhl, D-6900 Heidelberg 1, F.R.G

Abstract: Radio gaps, i.e. regions of strongly reduced radio (and also optical) intensity
between the galactic nucleus (AGN) and the radio jets in active radio galaxies are a
rather prominent feature of many of these sources (Bridle, 1986). With the aim to explain this phenomenon we consider the evolution of the relativistic electron spectrum in
such a source with the relevant plasma kinetic equations. We consider a possible acceleration process for relativistic electrons and positrons near the accreting central object
which is based on the continuous interplay of stochastic Fermi acceleration processes
and radiation losses of energetic particles (Schlickeiser, 1984). The resulting distribution function heats the background medium effectively through the oblique Langmuir
synchrotron instability (Lesch et al., 1989a) once the threshold plasma energy density
for the modulation instability is exceeded resulting in the outflow of material. By the
concerted action of synchrotron losses and adiabatic focussing in the ordered, spatially
decaying magnetic field B_0 the isotropic relativistic electron distribution function rapidly develops into to oppositely directed beam distributions, explaining the formation of
compact, superluminal radio jets near the AGN and the radio gap phenomenon. The
stability of the electron/positron beam in the magnetized background plasma is studied
by treating it as a cold beam. The linear growth rates for parrallel propagating longitudinal and transversal waves are calculated. Whereas the former give rise to heating of
the background plasma, the low-frequency transversal (Alfvén-) waves scatter the beam
to an isotropic distribution function which then reveals itself again by synchrotron radiation. A lower limit for the mean free path for this isotropization process is calculated
and compared with the length of the radio gaps. A discussion of the results is given.

1 Radiogaps

In many active radio galaxies there is a region of strongly reduced radio intensity between the AGN and the radio jet (Bridle et al., 1986). Supported by strong empirical evidence (Röser and Meisenheimer, 1986, Schlötelburg et al., 1988) we assume that the radiation is caused by the synchrotron process. Then a ready explanation for the gap phenomenon can be given by assuming that in the gap all particles move along the magnetic field lines, i.e. their distribution in momentum space is completely anisotropic. A gross picture of its development between AGN and radio jet is given in fig.1.

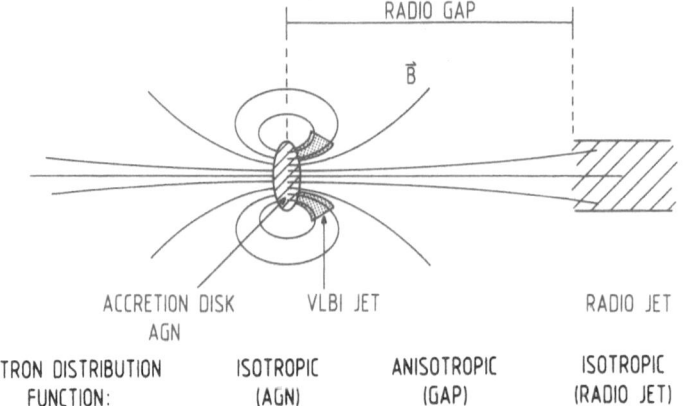

Fig. 1. Global picture of the VLBI-jet and the radio jet up to the first bright knot

In the following a theory will be presented with the aim to account for their behavior. Details are given in Achatz et al. (1990).

2 Particle Acceleration in the AGN

For the AGN we assumed the standard model of a supermassive accreting object (Rees, 1987). Matter is accreted through the galactic disk thereby leading to the formation of shocks (McCrea, 1956) and a poloidal magnetic field (Lesch et al., 1989b). In the whole active region Alfvén waves are present (Achatz et al., 1990). Therefore particle acceleration both by diffusive shock acceleration (Drury, 1983) and by quasilinear momentum diffusion (Schlickeiser, 1989a, 1989b) is possible which has to be balanced by loss processes like synchrotron and inverse Compton losses, Coulomb collisions and Bremsstrahlung (Schlickeiser and Crusius, 1988).

Schlickeiser (1984) has shown that the concerted action of these processes leads to a nearly monoenergetic distribution with a strong peak at the momentum where shock acceleration and synchrotron losses are balanced. It has furtheron been shown that this so-called pile-up is unstable against the excitation of oblique Langmuir waves (Lesch et al.,1989) which are first damped by nonlinear processes

but finally heat the background electron/proton plasma via the modulation instability strongly so that an explosive outward motion of the plasma results, carrying
the relativistic particles into the outer regions of the galaxy.

Since these are much more quiescent than the central part their Alfvén wave intensity should be decisively lower so that the sole influence the pile-up is exposed to
can be assumed to be from the magnetic field which we took to be poloidal (Lesch
et al., 1989b). Therefore the change of pitch angle and Lorentzfactor are caused by
adiabatic focussing in the spatially decaying magnetic field and synchrotron losses.
One can easliy check that the focussing of the pile-up can lead to the formation of
a nearly monoenergetic electron/positron beam propagating along the magnetic
lines. It has been shown by Sanders (1974) that this could give an explanation
for apparently superluminal VLBI-jets using the electrons moving not along the
polar axis. The remaining ones are never seen by the observer thereby giving an
explanation for the radio gaps (cf. fig. 1).

3 Isotropization of the Beam

In the frame of our model the finite length of the radio gap means that there has
to be some process which finally isotropizes the beam. Indeed it can be shown that
a monoenergetic electron/positron beam is unstable against the excitation of both
transversal and longitudinal waves propagating along the magnetic field which we
assume to be the fastest growing ones due to a result by Tademaru (1969). Since
experience from solar physics (Papadopoulos et al., 1974) and the physics of AGN's
(Lesch and Schlickeiser, 1987) has shown that nonlinear effects can lead to a very
efficient damping of Langmuir waves we assume that the transversal waves have
the most important effect on the beam.

In the case that the Alfvén velocity is much smaller than the velocity of light,
i.e.

$$\frac{v_A}{c} = 2.3 \ 10^{-3} \left(\frac{B_0}{10^{-4}\mathrm{G}}\right) \left(\frac{n_b}{0.1\mathrm{cm}^{-3}}\right)^{-0.5} \ll 1$$

where n_b is the density of the cold electron/proton background medium and B_0
the magnetic field strength, the action of these waves which will here be approximated by Alfvén waves on a spatially homogeneous distribution f_a of particles is
described by a pitch-angle diffusion equation (Schlickeiser, 1989a). The pitch-angle-
diffusion coefficient depends linearly on the intensities of forward and backward
moving waves. There time dependence is controlled by (Lee and Ip, 1987) linear
growth rates which depend again on the distribution functions of the electrons and
positrons so that the whole process, ending in the isotropization of the beam is
described by a system of four coupled equations. Assuming an weak initial wave
spectrum we find that the self-excited turbulence has a $|k|^{-2}$-dependence with an
amplitude

$$T = \frac{n_r}{n_b} \left(\frac{m_e}{m_p}\right)^{\frac{1}{2}} \frac{\omega_e}{c} B_0^2,$$

where n_r is the density of the relativistic beam and ω_e the plasma frequency of the background electrons.

Reinserting this turbulence into the diffusion equation we find a lower limit for the isotropization time τ. Multiplying this with the particle velocity one gets a length

$$l = c\tau = \frac{B_0^2}{\pi T} = 2.4\ 10^{13}\text{cm} \left(\frac{n_r}{10^{-7}\text{cm}^{-3}}\right)^{-1} \left(\frac{n_b}{0.1\text{cm}^{-3}}\right)^{\frac{1}{2}},$$

which is much less than 1 kpc, i.e. the typical gap length.

4 Conclusions

It was shown that by the concerted action of accretion, accretion shocks and Alfvén waves in and a poloidal magnetic field in the vicinity of an AGN the formation of monoenergetic electron/positron beams propagating along the magnetic field between the AGN and its radio jet is possible thereby giving a ready explanation for the well known radio gap phenomenon. An attempt to verify typical gap lengths has, however not yet been successful since so far we could only derive a lower limit for the typical isotropization length (10^{13}cm). More detailed calculations are necessary considering not only the corresponding time dependent problem but also including nonlinear wave damping effects like nonlinear Landau damping or the modulation instability of Alfvén waves.

References

Achatz, U., Lesch, H., Schlickeiser, R. 1990, *Astron.Astrophys.*, **233**, 233

Bridle, A.H. 1986, *Can.J.Phys.*, **64**, 353

Bridle, A.H., Perley, R.A., Henriksen, R.N. 1986 *Astron.J.*, **92**, 534

Drury, L.O.C. 1983, *Rep.Prog.Phys*, **46**, 973

Lee, M.A., Ip, W.H. 1987, *J.Geophys.Res.*, **92**, 11,041

Lesch, H., Schlickeiser, R. 1987, *Astron.Astrophys.*, **179**, 93

Lesch, H., Crusius, A., Schlickeiser, R. 1989a, *Astron.Astrophys.*, **209**, 427

Lesch, H., Crusius A., Schlickeiser, R., Wielebinski, R. 1989b, *Astron.Astrophys.*, **217**, 99

McCrea, W.N. 1956, *Ap.J.*, **124**, 461

Papadopoulos, K., Goldstein, M.L., Smith, R.A. 1974 *Ap.J.*, **190**, 175

Rees, M.J. 1987, "Jets and Galactic Nuclei" in *Highlights of Modern Astrophysics*, ed. by S.L. Shapiro and S.A. Teukolsky (Wiley, New York), pp. 163–190

Röser, H.J., Meisenheimer, K. 1986 *Astron.Astrophys.*, **154**, 15

Sanders, R.H. 1974, *Nature*, **248**, 390

Schlickeiser, R. 1984 *Astron.Astrophys.*, **136** 227

Schlickeiser, R. 1989a, *Ap.J.*, **336**, 231

Schlickeiser, R. 1989b, *Ap.J.*, **336**, 264

Schlickeiser, R., Crusius, A. 1988, *Ap.J.*, **328**, 578

Schlötelburg, M., Meisenheimer, K., Röser, H.J. 1988, *Astron.Astrophys.*, **202**, L23.

Tademaru, E. 1969, *Ap.J.*, **158**, 959.

Pion Production in Strong Magnetic Fields: A Model for Gamma-Ray Emission from Accreting X-Ray Pulsars

Charles D. Dermer

Department of Space Physics and Astronomy, Rice University
Houston, Texas 77251, USA

Abstract: Hadrons accelerated to high energies in the accretion column of a binary X-ray pulsar can produce pions in collisions with accreting protons and ions. The pions decay promptly into charged particles and photons, giving rise to an electromagnetic cascade through pair and photon production processes in the strong magnetic field of the neutron star. We give a simplified analysis of this system for teragauss magnetic fields in the optically-thin limit, assuming that nonlinear interactions between the produced pairs and photons are small, and calculate photon spectra in the MeV range. The calculated spectra are found to depend strongly on energy and angle, and are exponentially truncated at energies determined by the $\gamma - B$ pair production threshold, which is a function of the magnetic field strength and viewing angle with respect to the magnetic axis.

Gamma-ray observations of rotating magnetized neutron stars in the MeV range can provide a new method for determining (1) the magnetic field strength at the gamma-ray production site, and (2) the neutron star inclination angle. These possibilities are primarily due to $\gamma - B$ pair attenuation processes in the accretion column, and are not restricted to the specific pion production model for gamma-ray emission proposed here. The Gamma Ray Observatory, with unprecedented sensitivity at MeV energies, should be able to test these ideas.

1 Introduction

The existence of processes accelerating hadrons to relativistic energies in compact cosmic objects receives its strongest support from the detection of TeV gamma rays from point sources. This is because hadronic energy loss rates are much less than those of electrons, so energization of hadrons to high energies is thought to cease only when the time scale for escape from the system becomes comparable to the acceleration time scale, or when processes with high thresholds, such as photomeson production, begin to compete with the acceleration process. This picture fails, however, to take note of years of radio pulsar research involving electrons accelerated to Lorentz factors $\gamma \sim 10^{6-8}$ by the induced electric field of a spinning neutron star (see Sturrock 1971 and the review by Michel 1982), which suggests that directly accelerated electrons are actually responsible for the ultra-high-energy emission (Schlickeiser 1989).

In view of the theme of this workshop, I will nevertheless assume that observations of VHE gamma-ray emission do indeed signal the presence of highly relativistic hadrons. What guidance do observations give to the likely acceleration sites? According to Weekes (1989; see also Vacanti, these proceedings), the number of established sources in the VHE gamma-ray sky is just four: the Crab supernova remnant; the radio binary Cyg X-3, and the accreting X-ray binaries Her X-1 and Vela X-1. These sources probably all contain highly magnetized neutron stars. By contrast, no observation of $\gtrsim 10^{12}$ eV emission from black hole sources, whether galactic (e.g., Cyg X-1) or extragalactic (such as Cen A), has been confirmed. I therefore conclude that if TeV emission is taken as evidence for hadronic energization processes, then teragauss magnetic fields very likely play a decisive role in accelerating these particles.

Thus I consider additional potential radiative signatures of relativistic hadrons near a magnetized neutron star. In particular, I treat a system in which relativistic protons produce pions after colliding with low-energy protons in an intense magnetic field. The pions decay into gamma rays, electrons and positrons. Because the gamma-rays have energies $\gtrsim 10^{1-2}$ MeV, they promptly materialize into electron-positron pairs via magnetic pair production, provided that the angle between the gamma ray and magnetic field directions is sufficiently large. The electrons and positrons radiate energetic synchrotron photons that can likewise pair produce if they are sufficiently energetic. These successive generations of pair/photon production represent the synchrotron pair cascades studied recently by Baring (1989) and Preece and Harding (1989). Although $\gamma - \gamma$ pair production, resonant Compton scattering, and electron-positron annihilation could be important, we neglect these processes in the present analysis, and consider only the linear, optically thin case here.

In Section 2, I motivate this work by reference to a widely accepted model for X-ray pulsars. The well-known features of pion production in proton-proton collisions are summarized, and the modifications of the pion production and decay process due to the presence of a strong magnetic field are indicated. In Section 3, synchrotron emission and $\gamma - B$ pair production are reviewed. Approximate spectra in the optically thin regime are calculated in Section 4, showing that $\gamma - B$ attenuation governs the form of the observed spectra. This suggests a new method for measuring the magnetic field strength and inclination angle of an accreting X-ray pulsar, which should be testable with the Gamma Ray Observatory.

2 Pion Production in Accreting X-Ray Pulsars

Explanations for the pulsing X-ray emission observed from sources such as Her X-1, SMC X-1 and Cen X-3, have almost universally involved neutron stars in binary systems that accrete mass from their stellar companions either through Roche-lobe overflow or from stellar winds (see Joss and Rappaport 1984; Mészáros 1984; and Nagase 1989 for reviews). In the case of Her X-1, the phase-averaged X-ray spectrum displays an approximate power-law shape with spectral index $\alpha \approx 1$ between photon energies $\epsilon \sim 2$ and ~ 20 keV (see Figure 1), breaking to a very steep power law at $\epsilon \approx 20$ keV (White et al. 1983). At $\epsilon \gtrsim 30$ keV, features have been observed (Trümper et al. 1978; Mihara et al. 1990) that are thought to result from cyclotron absorption of photons by a plasma embedded in a magnetic field with $B \cong 2.9 \times 10^{12}$ Gauss. In the soft gamma-ray regime, no detections have been reported from Her X-1 or, for that matter, from any other binary X-ray pulsar. But because acceleration of hadrons to highly relativistic energies in X-ray pulsars may be the source of TeV gamma rays seen in binary X-ray pulsars (see Section 1), it seems likely that radiative signatures from these hadrons will also be seen in the MeV range, given the increased sensitivity possible with the telescopes on GRO.

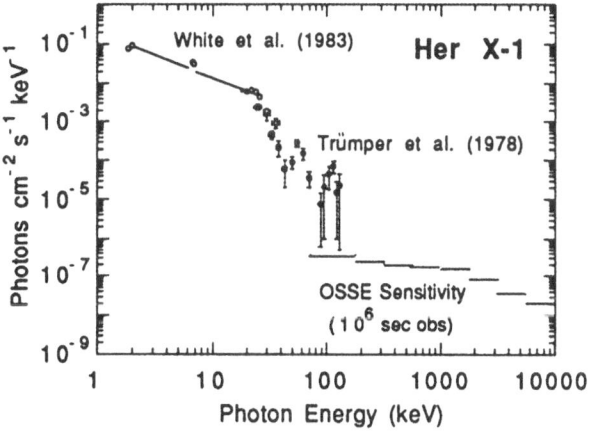

Fig. 1. Pulse-averaged spectrum of Her X-1, showing the presence of cyclotron features between 30 and 60 keV. The 3σ OSSE sensitivity (Kurfess et al. 1989) for a 10^6 second observation is also shown.

I therefore assume that there is a population of relativistic protons or ions in the accretion column of an accreting neutron star. This could result from a standing shock in the accretion stream, although the existence of such shocks has not been demonstrated (see Mészáros 1984). If there are protons with energies $E \gtrsim 300$ MeV, however, they can produce pions through collisions with particles in the accretion plasma. The fate of such pions, and the resultant gamma-ray signature, are now considered.

We first ask whether gamma-ray emission produced in the accretion column will escape freely, or will undergo a large number of Compton scatterings before escape. The radiative transport of gamma-ray photons through Compton scattering is elementary if the first condition is met. Let $A = \pi R^2$ denote the area of the accretion column, and let $\dot{N}$ denote the number of protons accreted per second. Thus $\dot{N} = L/\eta m_p c^2$, where $\eta \approx 0.1$ is the energy-conversion efficiency of matter falling through the gravitational potential of a neutron star. The average density of protons is given by $n_p \cong \dot{N}/A\beta_{ac}c$, where $\beta_{ac}c$ is the mean velocity of accreting matter at the gamma-ray production site. The Thomson depth through the accretion column, expressed in terms of the Eddington limit $L_{Edd} = 4\pi cGMm_p/\sigma_T$, where M is the neutron star mass and the other symbols have their usual meaning, is given by

$$\tau_T \cong \frac{f_{Edd}}{\beta_{ac}\eta} \left(\frac{4GM}{c^2 R} \right) \approx 8\frac{f_{Edd}}{\beta_{ac}\eta R_5}, \tag{1}$$

where $f_{Edd} = L/L_{Edd}$, and the accretion column radius $R = R_5 10^5$ cm. For a 1.4 solar mass neutron star radiating at $\approx 10\%$ of the Eddington luminosity, we find that the accretion column is therefore quite optically thick to Thomson scattering. Since we consider only the optically thin case here, we are therefore dealing with systems radiating at $\lesssim$ several percent of the Eddington luminosity. (Note that the optically thin criterion also depends on the photon energy through the Klein-Nishina decline in the Compton cross section.)

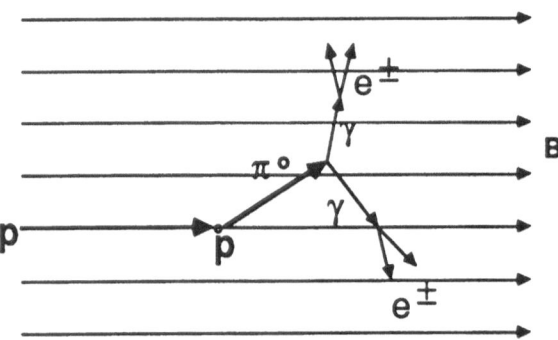

Fig. 2. Cartoon illustrating the collision between an energetic proton, travelling along the magnetic field direction, and a proton at rest. A neutral pion is produced and decays into two gamma-ray photons which materialize into pairs in the strong magnetic field region.

The elementary interaction is sketched in Fig. 2. A proton, travelling along a field line, produces a pion through a collision with a proton at rest. The energetic proton travels along the field line if it has radiated away most of its transverse momentum. This occurs if the collision time scale t_{p-p} is much longer than the synchroton energy-loss time scale t_{syn}. For a pion-producing event,

$$t_{p-p} \sim (n_p \sigma_{p-p} c)^{-1} \approx 10^{-5} R_5^2 \left(\frac{\eta \beta_{ac}}{f_{Edd}} \right) \text{sec,} \tag{2}$$

where $\sigma_{p-p} \approx 30$ mb at proton energies well above threshold ($\cong 300$ MeV). From the classical synchrotron energy-loss formula, one finds

$$t_{syn} \approx 5 \times 10^{-16} \left(\frac{m}{m_e} \right)^3 (\beta^2 \gamma B_{12}^2 \sin^2 \alpha)^{-1} \text{sec,} \tag{3}$$

where m stands for the particle mass, B_{12} is the magnetic field in teragauss, $\gamma = (1 - \beta^2)^{-1/2}$ is the Lorentz factor of the particle, and α is its pitch angle. For protons with $\alpha = \pi/2$ and $\gamma \gtrsim 2$, $t_{syn} \lesssim 2 \times 10^{-6} B_{12}^{-2}$. Comparing this result with equation (2), we see that it is usually a good approximation to assume that a proton radiates away most of its transverse energy before undergoing a pion-producing reaction.

Photons produced through neutral pion production in the reaction $p + p \to \pi^o \to 2\gamma$ can pair produce through $\gamma - B$ absorption in a strong magnetic field if the threshold condition

$$\epsilon \sin \alpha_\gamma \geq 2 \tag{4}$$

is met. Here ϵ is the dimensionless photon energy and α_γ is the photon pitch angle. Also, the characteristic attenuation length must be small compared to the size scale R of the system. These electrons and positrons will radiate synchrotron photons which can produce a new generation of pairs, thus initiating a synchrotron pair cascade (see next section).

The sequence of events in the reactions $p + p \to \pi^\pm \to \mu^\pm \to e^\pm$ involving charged pion production and decay depends on the relative decay and radiative time scales. For reference, we show in Table 1 the relevant particles, their masses, mass ratios (with respect to the electron mass), and mean lifetimes. Using equation (3), one can show that $t_{syn}^{\pi^\pm} \approx t_{decay}^{\pi^\pm}$, so that the charged pions may decay into muons before reaching low Landau levels. However, $t_{syn}^{\mu^\pm} \ll t_{decay}^{\mu^\pm}$. Thus the muons will be in their ground Landau levels and will be strongly polarized when they decay. The characteristic energy of synchrotron photons radiated by charged particles is given by

$$h\nu_B = \left(\frac{m_e}{m} \right) (e\hbar B/m_e c) \gamma^2 \cong \left(\frac{m_e}{m} \right) 11.6 \, \gamma^2 \, keV. \tag{5}$$

Electrons and positrons are clearly most important for gamma-ray production.

We also note that neutrons will be formed as a product in reactions like $p + p \to n + p + \pi^+$. The neutrons, which are not confined by the magnetic field of the neutron star, can escape from the compact object and form a neutron halo around the star. They could then interact with the companion star and produce either 2.2 MeV line radiation through proton capture deep in the stellar photosphere

TABLE 1
PARTICLE MASSES AND LIFETIMES[a]

Symbol	Mass (MeV)	m/m_e	Mean Lifetime (sec)	Branching Ratio %
π^0	134.963	264.1	0.83×10^{-16}	98.80
π^+, π^-	139.567	273.1	2.6030×10^{-8}	100.0
μ^+, μ^-	105.659	206.8	2.1971×10^{-6}	100.0
p	938.279	1836.1		

[a] Particle Data Group (1984).

(Dermer and Ramaty 1986; Vestrand 1989) or, if the energy of the secondary neutron is great enough, UHE gamma-ray emission (Kazanas and Ellison 1986).

3 Synchrotron and Pair Production Processes

Electron synchrotron radiation and magnetic pair production are the most important quantum electrodynamic processes in our problem (see Harding .1991 for a review of processes in strong magnetic fields). The treatment of the synchrotron process is complicated by the fact that both the classical synchrotron (CS) regime, with $\gamma\epsilon_B \ll 1$, and the quantum synchrotron (QS) regime, with $\gamma\epsilon_B \gg 1$, are encountered. Here $\epsilon_B = B/B_{cr}$, where $B_{cr} = 4.414\times10^{13}$ Gauss is the critical magnetic field (see Brainerd and Lamb 1987 for a more thorough discussion). Both regimes are important for electrons with $\gamma \gtrsim 10$ in a teragauss magnetic field.

The basic unit of time for electron synchrotron radiation is given by $\tilde{t}_{syn} = \hbar/\alpha_f m_e c^2 = 1.8 \times 10^{-19}$ sec. In units of $\tilde{t}_{syn}$, the asymptotic electron energy-loss rates in the two regimes are given by

$$|\dot{\gamma}| = \begin{cases} \frac{2}{3}(\epsilon_B\gamma)^2 \ , \ CS \\ \left(\frac{2}{3}\right)^5 (3\epsilon_B\gamma)^{2/3} \ , \ QS \end{cases} \tag{6}$$

the photon production rates by

$$\dot{N}_{ph} = \begin{cases} \frac{3^{1/2}\Gamma(1/6)\Gamma(11/6)}{2\pi}\epsilon_B \ , \ CS \\ \left(\frac{14}{27}\right)(3\epsilon_B)^{2/3}\gamma^{-1/3} \ , \ QS \end{cases} \tag{7}$$

and the average energies of synchrotron photons by

$$\bar{\epsilon} = \frac{|\dot{\gamma}|}{\dot{N}_{ph}} = \begin{cases} \frac{4\pi}{3^{3/2}\Gamma(1/6)\Gamma(11/6)}\epsilon_B\gamma^2 \cong 0.46\epsilon_B\gamma^2 \ , \ CS \\ \frac{16}{63}\gamma \ , \ QS \end{cases} \tag{8}$$

(see Brainerd and Petrosian 1987; Baring 1989). These expressions are valid for relativistic ($\gamma \gg 1$) electrons with pitch angle $\alpha = \pi/2$.

We can estimate the spectrum of synchrotron photons in the CS and QS limits using the δ-function approximation

$$\dot{n}_{ph}(\epsilon) = \int_{1+\epsilon}^{\infty} d\gamma \, n_e(\gamma) \, \dot{N}_{ph}(\gamma, \epsilon_B) \, \delta[\epsilon - \bar{\epsilon}(\gamma, \epsilon_B)], \tag{9}$$

where the steady-state electron distribution function $n_e(\gamma) = |\dot{\gamma}|^{-1} \int_\gamma^\infty d\gamma' \dot{n}_e(\gamma')$, and $\dot{n}_e(\gamma)$ represents the electron injection function. Reinjection of electron-positron pairs through $\gamma - B$ pair production can be approximated by

$$\dot{n}_e^r(\gamma) = 2 \int_{\epsilon_{co}}^{\infty} d\epsilon \, \dot{n}_{ph}(\epsilon) \, \delta[\gamma - \epsilon/2]. \tag{10}$$

The $\gamma - B$ attenuation cutoff energy ϵ_{co} is obtained from the absorption coefficient

$$\kappa_{\gamma-B}[\mathrm{cm}^{-1}] = \frac{\epsilon_B \sin \alpha_\gamma}{2 a_B} \, \mathrm{T}(\chi) \, \mathrm{H}(\epsilon \sin \alpha_\gamma - 2) \tag{11}$$

(Erber 1966; see also Ho et al. 1990), where a_B is the Bohr radius, $\chi = \epsilon \, \epsilon_B \sin \alpha_\gamma/2$, and the H-function equals 1 and 0 for positive and negative arguments, respectively. When $\chi \ll 5$, $T(\chi) \to 0.46 \exp(-4/3\chi)$, and in the limit $\chi \gg 5$, $T(\chi) \to 0.6\chi^{-1/3}$. Photons with energies near threshold will be attenuated in a distance of $10^5 R_5$ cm if

$$\epsilon \gtrsim \epsilon_{thr} = \frac{4.7}{\sin \alpha_\gamma B_{12}[1 + 0.0435 \ln(B_{12} R_5 \sin \alpha_\gamma)]} \geq \frac{2}{\sin \alpha_\gamma}. \tag{12}$$

In equation (10), the cutoff photon energy $\epsilon_{co} = \epsilon_{thr}(\alpha_\gamma = \pi/2)$.

We now describe the sequence of events encountered in a synchrotron pair cascade (see Baring 1989 for a detailed treatment). This involves electron injection, synchrotron radiation, $\gamma - B$ photon attenuation, and pair reinjection. The reinjected pairs radiate additional photons, and the cascade terminates when the energy of the synchrotron photons fall below the energy given by equation (12). The cascade is most easily studied in the frame in which the primary photon has pitch angle $\alpha_\gamma = \pi/2$. We illustrate the development of the cascade using equations (9) and (10):

(i) Let the electron injection function $\dot{n}_e(\gamma) \propto \gamma^{-\Gamma}$;

(ii) Then $n_e(\gamma) \propto \gamma^{-p}$, where $p = \Gamma + 1$ (CS), and $p = \Gamma - 1/3$ (QS);

(iii) The photon spectrum $n_{ph}(\epsilon) \propto \epsilon^{-s}$, with $s = 1 + \Gamma/2$ (CS), and $s = \Gamma$ (QS);

(iv) The electron reinjection function $\dot{n}_e(\gamma) \propto \gamma^{-\Gamma^r}$, where $\Gamma^r = s = 1 + \Gamma/2$ (CS) and $\Gamma^r = s = \Gamma$ (QS).

Thus we find that the reinjected electron-positron pair spectrum in the CS limit has a different spectral index than that of original injection function (unless $\Gamma = 2$). But for teraguass magnetic fields and an electron injection function extending into the QS regime, only a small fraction of the reinjected pairs are formed

by synchrotron photons produced in the CS regime. The bulk are made by magnetic pair production of synchrotron photons produced in the QS regime. In this
regime, the spectral index of the reinjection function equals the original injection
spectral index Γ. Thus the pair cascade does not alter the classical spectral index
of the photon synchrotron spectrum produced by a cooling electron distribution,
but does increase its luminosity (Baring 1989). One can therefore approximately
describe the photon spectrum produced in a synchrotron pair cascade using classical synchrotron formulae, but with a spectrum that is truncated at the $\gamma - B$
cutoff energy ϵ_{co} in the frame in which $\alpha_\gamma = \pi/2$. Moreover, the intensity of the
spectrum is enhanced by an amount proportional to the total injected energy of
electrons with $\gamma > \epsilon_{co}$.

4 Pion Production in Strong Magnetic Fields

In this section, we give a somewhat intuitive derivation of the photon spectrum
produced through pion production in strong magnetic fields. For simplicity, we
consider neutral pion production only. The kinematics of electrons and positron-
s produced in charged-pion decay are, however, similar to those of the neutral
pion-decay case, so we expect the cascade gamma-ray spectra formed through
synchrotron and pair proceses in the two cases to be similar.

In parallel with the treatment of secondary pion production in cosmic-ray proton collisions (cf. Dermer 1986), we write the angle-dependent production spectrum of neutral pions as

$$\dot{n}_\pi(\gamma_\pi, \mu_\pi) = n_p \int d\gamma_p \int d\Omega_p \, j_p(\gamma_p, \Omega_p) \frac{d^2\sigma(\gamma_\pi, \mu_\pi; \gamma_p)}{d\gamma_\pi d\mu_\pi} , \qquad (13)$$

where n_p is the density of low-energy protons, and the one-dimensional proton flux

$$j_p(\gamma_p, \Omega_p) \, [\text{protons cm}^{-2}\text{s}^{-1}\text{sr}^{-1}] = \frac{1}{2} \, \Phi(\gamma_p) \, [\delta(\Omega_p - 1) + \delta(\Omega_p + 1)] . \qquad (14)$$

We employ a crude order-of-magnitude estimate for the π^0 differential cross section. Recognizing that the most energetic pions are made through isobar formation
and will therefore be formed with Lorentz factor $\approx \gamma_p$, and noting moreover that
the spectral index of secondary π^0 is equal to that of the primary protons in the
extreme relativistic limit, we have

$$\frac{d^2\sigma(\gamma_\pi, \mu_\pi; \gamma_p)}{d\gamma_\pi d\mu_\pi} \sim \sigma_{\pi^0} \, \delta(\gamma_\pi - \gamma_p)\delta(\mu_\pi - 1) , \text{ for } \gamma_p \geq \gamma_{p,min} \approx 2 . \qquad (15)$$

Here we also assume that the pions are beamed in the direction of the incident
proton ($\mu_\pi = 1$). This is valid for extremely energetic collisions when $\gamma_p \gg 1$. At
lower energies, the angular distribution of the pions is quite broad. But in this
case, $\gamma_\pi \sim 2$, and the angular distribution of pion-decay gamma rays is not much
different for either beamed or isotropic pions. To agree with the magnitude of

the secondary π^o flux in cosmic-ray collisions when $\gamma_p, \gamma_\pi \gg 1$, $\sigma_{\pi^o} \approx 25$ mb in equation (15).

Let $\epsilon_\pi = m_{\pi^o}/2m_e \cong 132$ represent the dimensionless gamma-ray photon energy emitted in the rest frame of a decaying neutral pion. Since two photons are made per decay, the distribution of photons in the observer's frame is given by

$$n_{ph}(\epsilon,\mu)d\epsilon d\mu = 2 \times \frac{1}{2}\delta(\epsilon' - \epsilon_\pi)d\epsilon'd\mu'. \tag{16}$$

The Lorentz transformation equations give

$$n_{ph}(\epsilon,\mu) = \frac{\epsilon}{\epsilon_\pi}\delta[\gamma_\pi\epsilon(1 - \beta_\pi\mu) - \epsilon_\pi]. \tag{17}$$

Using the previous approximations, we find that the angle-dependent production spectrum of cascade gamma rays from the decay of pions produced in proton-proton collisions is given by

$$\dot{n}_{ph}(\epsilon,\mu) = n_p\sigma_{\pi^o}\frac{\epsilon}{\epsilon_\pi}\int_{\gamma_{\pi,min}}^{\infty} d\gamma_\pi\ \Phi(\gamma_\pi)\ \delta[\gamma_\pi\epsilon(1 - \beta_\pi\mu) - \epsilon_\pi]\ . \tag{18}$$

In this expression, we consider only one direction of the protons' motion.

As described in the preceeding section, a pair cascade will result from $\gamma - B$ attenuation of pion-decay gamma rays. To determine the spectrum seen by an observer, we need to consider transformations between three reference frames. Gamma rays from π^o decay are formed with energy ϵ_π in the rest frame (RS) of the pion. The cascade is most easily treated in the pair cascade (PC) frame where the photon pitch angle $\alpha_\gamma = \pi/2$, or $\mu'' = 0$. The observed gamma-ray spectrum is calculated in the neutron star (NS) frame. Note that: (i) The relationship between μ, ϵ, and ϵ_π imposed by the δ-function in equation (18) shows that most pion-decay gamma rays are energetic enough to pair produce in teragauss magnetic fields (compare threshold condition [11]). (ii) The shower of secondary photons produced in the pair cascade will have a small angular spread about the direction of the primary pion-decay gamma ray, since the pairs are highly relativistic. Thus the cascade photons can be treated as having $\mu'' = 0$. (iii) Because nearly all the primary photons have energies $\epsilon'' \gg \epsilon_{co}$, the cascade gamma-ray spectrum can be approximated by a quasi-monoenergetic photon source function, implying that the PC photon spectrum $n''_{ph} \propto \epsilon''^{-3/2}$. (iv) The Lorentz transformation $\mu'' = (\mu - \beta_t)/(1 - \beta_t\mu)$ between the PC and NS frames shows that all photons produced at $\mu'' = \pi/2$ in the PC frame are directed at an angle $\mu = \beta_t$ in the NS frame, where $\beta_t c$ is the transformation velocity between the two frames.

The spectrum seen in the NS frame thus represents a transformation of the $\epsilon''^{-3/2}$ spectrum formed at $\mu'' = \pi/2$ in the PC frame. Transforming the angle-dependent production spectrum $\dot{n}''_{ph}(\epsilon'',\mu'')d\epsilon''d\mu''dt''$ from the PC to the NS frame involves an expression akin to equation (16), where

$$\dot{n}''_{ph}(\epsilon'',\mu'' = 0) = K(\mu)\ \epsilon''^{-3/2}H[\epsilon_{co} - \epsilon''] \tag{19}$$

and $K(\mu)$ is a normalization constant, which is a function of μ or β_t only. The Lorentz equations give $dt = \gamma_t dt''$ and $\epsilon'' = \epsilon/\gamma_t$, where $\gamma_t = (1 - \mu^2)^{-1/2}$. Thus

$$\dot{n}_{ph}(\epsilon, \mu) = K(\mu) \left(\frac{\epsilon}{\gamma_t}\right)^{-3/2} H[\epsilon_{co} - \epsilon/\gamma_t]. \tag{20}$$

The integrated photon energy emitted in the direction μ in the NS frame is obtained by weighting equation (20) by ϵ and integrating over ϵ, and using the H-function to limit the integration. One obtains

$$\frac{dE}{dV\,dt\,d\mu} = \frac{2\,K(\mu)\epsilon_{co}^{1/2}}{(1 - \mu^2)}. \tag{21}$$

The pion production spectrum (18) establishes the value of $K(\mu)$, since the integrated photon energy is also given by

$$\frac{dE}{dV\,dt\,d\mu} = n_p \sigma_{\pi^0} \epsilon_\pi \int_{\gamma_{\pi,min}}^{\infty} \frac{\Phi(\gamma_\pi)}{[\gamma_\pi(1 - \beta_\pi\mu)]^3}. \tag{22}$$

If the pion flux $\Phi(\gamma_\pi) \propto \gamma_\pi^{-r}$, then

$$K(\mu) \propto (1 - \mu^2) \int_{\gamma_{\pi,min}}^{\infty} \frac{\gamma_\pi^{-(r+3)}}{(1 - \beta_\pi\mu)^3}. \tag{23}$$

Equations (20) and (23) represent the principal results of this work. They show that the spectrum resulting from a synchrotron pair cascade initiated by pion-decay gamma rays in a strong magnetic field has a photon spectral index s $\cong$ 3/2, the same as a classical synchrotron cooling spectrum. The observed spectrum is cutoff at photon energies $\epsilon \gtrsim \epsilon_{co}/(1-\mu^2)^{1/2}$, where ϵ_{co} is obtained from equation (12) with $\alpha_\gamma = \pi/2$. The overall intensity of the spectrum is a strong function of the viewing angle, as indicated by equation (23). Our conclusions appear to agree with the numerical simulations of Preece and Harding (1989) for the cases they study in which isotropic relativistic electrons are injected into a teragauss magnetic field region. There they find that the angle-averaged cascade photon spectra have s $\approx$ 3/2 between the cyclotron energy and pair production cutoff. Our system should approximate these cases, since most of the secondary pions have small γ_π, and the pion-decay photons will therefore be injected nearly isotropically.

Figure 3 shows gamma-ray spectra calculated from equations (20) and (23) as a function of observing angle. Here we take $B_{12} = 2$, r = 2 and $R_5 = 1$. The sharp features in the spectra are artifacts of the analytic treatment. As can be seen, both the cutoff energy and intensity are strongly dependent on viewing angle. The discovery of a rotational phase variation in the cutoff energy of X-ray pulsar gamma-ray spectra would also reveal important information about the relative orientation the neutron star with respect to the observer. We encourage a search for such an effect. Because a large number of approximations were made in this treatment, however, comparison of these calculations with anticipated observations of X-ray binaries must be made with some caution. As discussed earlier, we have assumed that the system was optically thin to Compton scattering. The resonances in the Compton cross section (Daugherty and Harding 1986; Dermer 1990) at the cyclotron fundamental and harmonics may increase the importance of this effect. Neither have we considered the attenuation of synchrotron gamma-rays by $\gamma - \gamma$ pair production with other synchrotron photons. Moreover, the cascade photons could be absorbed by annihilation radiation made as the pairs cool. In a luminous X-ray source such as Her X-1, the X-ray photon field could also attenuate the gamma-ray emission (A. Atoyan, private communication, 1990). The inclusion of these nonlinear effects requires a detailed pair-balance calculation (Harding 1984).

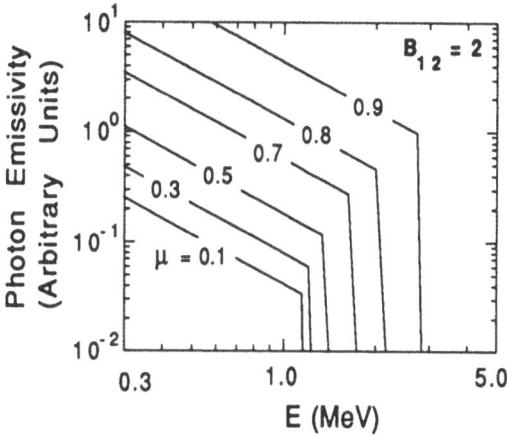

Fig. 3. Angle- and energy-dependent gamma-ray spectra resulting from neutral pion production and decay in a strong magnetic field. The angle between the photon and magnetic field directions is denoted by $\theta = \arccos \mu$.

In spite of the preliminary nature of this analysis, some interesting general conclusions follow:

- Gamma rays formed in the accretion column of a neutron star are subject to $\gamma - B$ pair attenuation, independent of the nature of their origin. The dependence of the cutoff energy with rotational phase depends on strength of the magnetic field and the orientation of the observer. Thus gamma-ray observations of accreting binary X-ray pulsars in the MeV range could provide a new method to determine the magnetic field strength and inclination axis of an accreting neutron star. (These considerations may also apply to gamma rays from radio pulsars, but pulsed TeV gamma rays from the Crab suggest that the emission site is not in the vicinity of the polar cap.)

- Complementary measurements of the magnetic field strength from observations of cyclotron line features and through the method proposed here can be used to determine the relative locations of X-ray and gamma-ray production sites in the accretion column of a neutron star.

• In the linear, optically thin limit, the spectral index of gamma rays formed from synchrotron pair cascades of pion-decay gamma rays is $\approx 3/2$ at energies below the $\gamma - B$ attenuation cutoff, independent of phase. Observations of such a spectrum would support arguments for the existence of relativistic hadrons in the accretion column of accreting X-ray pulsars. Other spectral indices may tend to argue in favor of directly accelerated electrons. However, definitive conclusions must await more detailed calculations.

In summary, a new method has been proposed for measuring magnetic fields and inclination angles of accreting X-ray pulsars, based on the rotational phase dependence of the magnetic pair production cutoff at gamma-ray energies. The Oriented Scintillation Spectrometer Experiment onboard the Gamma Ray Observatory, with an order-of-magnitude greater sensitivity than previous gamma-ray telescopes, has the potential to test this method.

Acknowledgements: I would like to thank the organizers, especially Andrzej Zdziarski, for the opportunity to participate in the workshop. I also thank A. M. Atoyan for comments on this work, and Jon Arons and Dick Lamb for stimulating my interest in this subject.

References

Baring, M. G.: 1989, *Astr. Ap.* **225**, 260

Brainerd, J. J., and Lamb, D. Q.: 1987, *Ap. J.* **313**, 231

Brainerd, J. J., and Petrosian, V.: 1987, *Ap. J.* **320**, 703

Daugherty, J. K., and Harding, A. K.: 1986, *Ap. J.* **309**, 362

Dermer, C. D., and Ramaty, R.: 1986, in *Accretion Processes in Astrophysics*, eds. J. Audouze and J. Tran Thanh Van (Editions Frontieres: Singapore), p. 85

Dermer, C. D.: 1986, *Astron. Ap.* **157**, 223

Dermer, C. D.: 1990, *Ap. J.* **360**, 197

Erber, T.: 1966, *Rev. Mod. Phys.* **38**, 626

Harding, A. K.: 1984, in *Proc. of Varenna Workshop on Plasma Astrophysics* (ESA SP-207), p. 205

Harding, A. K.: 1991, *Phys. Rept.*, in press

Ho, C., Epstein, R.I., and Fenimore, E.E.: 1990, *Ap. J. Lett.* **348**, L25

Joss, P. C.:, and Rappaport, S. A.: 1984, *Ann. Rev. Astr. Ap.* **22**, 537

Kazanas, D., and Ellison, D. E.: 1986, *Nature* **319**, 380

Kurfess, J.D., et al.: 1989, in *Proc. of the Gamma Ray Observatory Science Workshop*, ed. W. N. Johnson, p. 3-35

Mészáros, P.: 1984, *Space Sci. Rev.* **38**, 325

Michel, F. C.: 1982, *Rev. Mod. Phys.* **54**, 1

Mihara, T., et al.: 1990, *Nature* **346**, 250

Nagase, F.: 1989, *Publ. Astr. Soc. Japan* **41**, 1

Particle Data Group: 1984, *Rev. Mod. Phys.* **56**

Preece, R., and Harding, A. K.: 1989, *Ap. J.* **347**, 1128

Schlickeiser, R.: 1989, *Astr. Ap.* **213**, L23

Sturrock, P. A.: 1971, *Ap. J.* **164**, 529

Trümper, J. et al.: 1978, *Ap. J. Lett.* **219**, L105

Vestrand, W. T.: 1989, in *Proc. of the Gamma Ray Observatory Science Workshop*, ed.
   W. N. Johnson, p. 4-274

Weekes, T. C.: 1989, in *Proc. of the Workshop on Astrophysics in Antarctica*, ed. D. J.
   Mullan, et al. (AIP: New York), p. 3

White, N. E., Swank, J. H., and Holt, S. S.: 1983, *Ap. J.* **270**, 711

Particle Acceleration at Shock Fronts in AGNs and Jets

John G. Kirk

Max-Planck-Institut für Kernphysik, Saupfercheckweg 1
6900 Heidelberg 1, Federal Republic of Germany

Abstract: Several aspects of the theory of first order Fermi acceleration at shock fronts are discussed which may be relevant for active galactic nuclei (AGNs) and jets. In the test particle picture, recent work on acceleration at oblique shocks is reviewed which indicates that such shocks produce a harder spectrum than parallel shocks. They also tend to accumulate a substantial excess of energetic particles ahead of the front (snow plough effect). Next, models predicting the maximum energy to which electrons can be accelerated at hot spots in jets are discussed. These models are successful in explaining the observed synchrotron spectrum, but operate only if the Alfvén turbulence on each side of the shock front is of a very special nature. Finally, some recent work on the nonlinear problem of the back reaction of accelerated particles on a relativistic fluid flow is summarized, which indicates that higher acceleration efficiencies may be achieved in relativistic shocks than can be obtained in nonrelativistic shocks.

1 Introduction

Particle acceleration mechanisms can be divided into two kinds: stochastic and deterministic. At shock fronts, an example of the latter is the so-called 'shock-drift' mechanism, in which particles are assumed to follow well-determined trajectories according to the Lorentz equation. On encountering a shock front, the trajectory may be reflected or transmitted, according to the phase and pitch angle of the encounter. Depending on the reference frame in which events are observed, the particle may lose or gain energy as a result of this interaction (Webb, Axford and Terasawa 1983). However, the energy change is predetermined once the initial conditions are specified, so that the acceleration is a 'one off' process and cannot proceed to arbitrarily high energy. Although this mechanism may be important in producing the enhancements of surface brightness seen in the hot spots of radio jets (Begelman and Kirk 1990) it is unlikely play a role in the acceleration of relativistic hadrons, which is the main topic of this workshop. For this problem, it is certainly necessary to invoke an acceleration mechanism capable of increasing a particle's energy over many decades. A stochastic mechanism is best suited to this demand, since there is no inherent limit to the energy attainable by at least some particles. Only when loss processes or the effects of the finite system geometry are included does an upper bound on energy arise. Especially in the case of hadrons, this upper bound can be very high indeed (Hillas 1984).

At shock fronts, the most important stochastic acceleration process is a first-order Fermi one between scattering centres moving with roughly the same speed as the fluid. In the case of nonrelativistic fluid motion, this process is called 'diffusive acceleration' because the particle transport can be described by a diffusion equation (for reviews see Drury 1983, Forman and Webb 1985, Blandford and Eichler 1987). Relativistic motion demands a more elaborate treatment of the anisotropy of the accelerated particles. In both cases, however, the basic assumptions about the physical conditions are the same:

1. Particles may be divided into those constituting the background plasma (the 'thermal component') and those which undergo acceleration (the 'nonthermal component').
2. At a shock front, the kinetic energy of the thermal component is converted into internal energy by some collisionless process, producing a converging flow pattern.
3. The nonthermal particles undergo elastic scattering off fluctuations in the magnetic field which are essentially anchored in the thermal component.
4. At the shock front, the nonthermal particles suffer no interaction with the electric and magnetic fields responsible for the collisionless relaxation of the thermal component.

Given these conditions, it is possible to calculate the steady state spectrum of nonthermal particles, assuming these behave as test particles. In the nonrelativistic case, the general solution of the cosmic ray (CR) transport equation can be written down immediately for both the upstream and downstream regions. The matching of particle density and streaming across the shock front then supplies the power law index (see, for example Drury 1983). In the relativistic case, the full angular dependent transport equation must be used. The general solution in the upstream and downstream regions can then be represented by eigenfunction expansions and the matching of the distribution function at the shock front again gives the required power law index (Kirk and Schneider 1987).

2 Oblique Shocks

Interestingly, the simple result of nonrelativistic theory applies both to parallel and oblique shock fronts, despite the fact that the two types of shock are quite different. Oblique shocks can reflect particle trajectories, whereas parallel ones cannot, and reflection is an effective way of driving the nonthermal particles into an anisotropic distribution. The process of first order Fermi acceleration is, therefore, strongly affected, provided the scattering is not sufficiently strong to supress this anisotropy.

At very low plasma speeds, scattering will always win, but reflection starts to play a role when the velocity of the point of intersection of a magnetic field line and the shock front v_{int} moves with a speed which is not small compared to the speed of the nonthermal particles (usually close to the speed of light). To determine when this occurs, one must, of course, specify the reference frame in which the speed is to be measured. It turns out (Kirk and Heavens 1989) that the correct frame is that in which the plasma motion is along the ambient magnetic field direction (de Hoffmann and Teller 1950). Not all shocks admit a transformation to such a de Hoffmann-Teller (dHT) frame: if the speed v_{int} is superluminal in any frame of reference, then it is superluminal in all, and there exists no dHT frame. In this case it is possible to transform to a frame in which both the electric and magnetic fields lie in the plane of the shock, i.e., the shock is exactly perpendicular. Trajectories can then no longer be reflected, and particles cannot recross the shock from downstream to upstream unless there is effective cross-field transport. These shocks will be called 'superluminal'. It is to be expected that first order Fermi acceleration is supressed at them.

In the limit of no cross-field transport, a semi-analytic treatment of the oblique (subluminal) shock problem is given by Kirk and Heavens (1989). The most important result is that reflection hardens the steady state particle spectrum. This can be understood by noting that first order Fermi acceleration at shocks depends on two quantities: the average gain in energy on performing either a reflection or a crossing and recrossing of the shock front, together with the probability of being swept away from the acceleration zone by the downstream flow. Now, as reflection becomes more and more important, i.e., as v_{int} increases, the average energy gain on reflection increases without limit. However, the phase space region containing particles which penetrate downstream and can, therefore, be swept away, does not expand to cover all incident particles, but remains roughly within a loss-cone of opening angle $\arccos(\sqrt{1 - 1/r})$, where r is the compression ratio of the shock front. The net result is that the steady state spectrum tends to p^{-3} (where p is the particle momentum), as $v_{int} \to c$.

Recently, Monte Carlo methods have been applied to simulate this problem (Ostrowski 1991). Without cross-field transport, the most interesting result is that high energy particles do indeed pile up ahead of the shock front as a result of reflections, leading to a substantial enhancement of the density in the upstream region. This phenomenon can aptly be termed the 'snow plough' effect; not only are particles prevented from crossing the shock front, they are also simultaneously moved sideways along it by the magnetic field, an effect first pointed out by Jokipii (1982). On inclusion of cross field scattering, the effect diminishes. In fact, for strong cross-field scattering, the direction of the magnetic field ceases to be important, and the results derived for equivalent parallel shocks are recovered.

3 Saturation

Were it not for loss processes, the steady state spectrum produced by the first order Fermi process at a shock front would be a power-law extending up to infinite energy. At some stage, it is clear that processes such as synchrotron radiation or nuclear collisions will become important and limit the energy achieved. To estimate just what the highest energy is, it is sufficient to estimate the point at which the acceleration rate roughly equals the loss rate. Let us turn first of all to the acceleration rate.

In the case of nonrelativistic flows, the distribution function is almost isotropic. The average time required for a particle crossing into the upstream region to return to the shock is

$$\Delta t = \frac{4\kappa}{|u_- |v},\tag{1}$$

where κ is the spatial diffusion coefficient of the nonthermal particles, u_- is the speed of the plasma into the shock front and v is speed of the nonthermal particles (in most cases $v \approx c$). An interesting property of this expression is that it is independent of the sign of the plasma velocity. Consequently, the average time taken for a particle leaving the shock into the downstream region to return to it is given by the same expression, except that the downstream plasma speed u_+ must be subsituted for u_-. At first sight, this result runs counter to one's intuition: particles which are swept back towards the shock front (i.e., those in the upstream region) must surely take less time to return than those which are always being swept away (i.e., those in the downstream region). Not true! In fact, the only difference between downstream and upstream is that some particles escape on the downstream side. But, if the plasma velocities are the same on each side, the average time for an excursion of those particles which *do* return is the same on each side of the shock. In their Monte-Carlo simulations, Ellison, Jones and Reynolds (1990) were able to exploit this property.

To find the acceleration time, one must combine the cycle time for shock crossing and recrossing with the average gain per cycle. Using nonrelativistic kinematics, the average momentum gain for an isotropic particle distribution is

$$\frac{\Delta p}{p} = \frac{4}{3}\frac{\Delta u}{v},\tag{2}$$

where Δu is the difference between the upstream fluid speed u_- and the downstream fluid speed u_+. We can now write the acceleration rate:

$$t_{\mathrm{acc}}^{-1} = \frac{\Delta u}{3\kappa}\frac{u_- u_+}{u_- + u_+}\tag{3}$$

If we employ the frequently used assumption of Bohm diffusion, $\kappa = r_g v/3$, where r_g is the particle gyro radius, this expression can be written in terms of the gyrofrequency Ω of the nonthermal particle as:

$$t_{\mathrm{acc}}^{-1} = \frac{\Delta u}{u_- + u_+}\frac{u_- u_+}{v^2}\Omega.\tag{4}$$

For nonrelativistic flows, therefore, the acceleration rate is slower than gyration about the magnetic field by a factor of the order of $O\left(u^2/v^2\right)$ (Hillas 1984). It is interesting to note that (4) leads to roughly the same acceleration rate a particle would have if it were able to tap directly (e.g., by the shock drift mechanism) the electric field induced in a conducting plasma moving with speed $\sim u_-$ through a magnetic field. For relativistic flows, one might expect a faster rate, but in this case (4) is invalid, since it relies not only on diffusive transport, but also on nonrelativistic kinematics. The only calculations to date to address this problem are Monte-Carlo simulations (Ellison, Jones and Reynolds 1990, Quenby and Lieu 1989). Results range from a factor 3 to 13 faster than the rate given by (4). Thus, although first order Fermi acceleration at a relativistic shock front is not strictly determined by (4), it would seem from the simulations that the acceleration rate does not exceed it by a large factor. Oblique shocks have also been suggested as possible sites of rapid acceleration (Jokipii 1988, Ostrowski 1991), but they can be regarded as a mixture of shock-drift and first order Fermi acceleration (Jokipii 1982, Kirk and Heavens 1989) and so are unlikely to exceed substantially an acceleration rate equal to Ω.

Of the various loss processes, two are likely to be important in the case in which the nonthermal particles are electrons. These are synchrotron radiation and inverse Compton scattering, for which the energy loss rates are

$$t_{\text{sync}}^{-1} = 3 \cdot 2 \times 10^{-8} \gamma \left(B^2/8\pi\right) \text{ secs}^{-1} \qquad (5)$$

(with B expressed in gauss, and $\gamma = (1 - v^2/c^2)^{-1/2}$) for synchrotron radiation and the same expression with the magnetic field energy density replaced by the energy density in the photon field U for inverse Compton scattering:

$$t_{\text{compt}}^{-1} = 3 \cdot 2 \times 10^{-8} \gamma U \text{ secs}^{-1} \qquad (6)$$

These loss processes apply to charged hadrons too, but the rates are reduced by the factor $(m/M)^3 \approx 1 \cdot 6 \times 10^{-10}$. More important for protons are p–p collisions:

$$t_{\text{pp}}^{-1} \approx 6.7 \times 10^{-16} n \text{ secs}^{-1}, \qquad (7)$$

(where n is the background density of protons in cm^{-3}) as well as processes involving protons colliding with photons. Most interesting among these are electron/positron pair production and photo-pion production. The former process has a much lower threshold – about 1 MeV, compared with about 145 MeV for pion production – and so would usually have more target photons. However, it also has a much smaller inelasticity – 1/1000 compared with about 1/2 for pion production. Which of the two dominates depends on the spectrum of photons present, but for the simple case of a power law spectrum with intensity inversely proportional to frequency, they turn out to be almost equally effective loss processes (Sikora et al. 1987):

$$t_{\text{pair}}^{-1} \approx t_{\text{pion}}^{-1} \approx 3 \times 10^{-15} \gamma \bar{U} \text{ secs}^{-1}, \qquad (8)$$

where $\bar{U}$ is the photon energy density per frequency decade. Comparing (6) and (8), assuming U and $\bar{U}$ are not vastly different from each other, leads to the conclusion that inverse Compton scattering is negligible for protons. The same conclusion applies to synchrotron losses too, unless the magnetic energy density exceeds that in photons by more than three orders of magnitude.

The maximum energy can now be calculated by equating the total loss rate to the acceleration rate. Given Bohm diffusion, one finds a maximum momentum $p_{max} \propto B^{-1/2}$. An interesting corollary of this is that the typical frequency ν_{max} of synchrotron photons emitted by electrons of the maximum energy is *independent* of the magnetic field, provided it is the synchrotron radiation itself which dominates the loss rate. One then obtains:

$$\nu_{max} = \frac{27}{4} \left[\frac{\hbar c}{e^2} \right] \left[\frac{mc^2}{h} \right] \left[\frac{u_- u_+ \Delta u}{v^2 (u_- + u_+)} \right] \tag{9}$$

In this expression, the first term in brackets is just the reciprocal of the fine structure constant. The second establishes that the frequency scale is the electron rest mass and the third is a dimensionless factor typically $O\left(u_-^2/c^2\right)$. Thus, for Bohm diffusion, we would expect to see synchrotron photons from shock accelerated electrons up to a cut off at about $500 \times (u_-/c)^2$ MeV. In several sources a cut-off in the synchrotron spectrum has been observed at about 3×10^{14} Hz, (Röser 1989), which is compatible with (9) only if the shock front responsible is extremely slow (≈ 10 km/sec). These sources are hot spots in the jets of radio galaxies of the Fanaroff and Riley type II, for which there is strong evidence that the bulk motion of fluid in the jets is at least mildly relativistic (e.g., Laing 1988). It appears, therefore, that either synchrotron losses are not the factor determining the highest electron energy, or that Bohm diffusion is a very poor approximation.

Biermann and Strittmatter (1987) and Roland, Pelletier and Muxlow (1988) have proposed models which overcome the above difficulty by postulating a different momentum dependence of the diffusion coefficient. Biermann and Strittmatter assume the turbulence which is ultimately responsible for diffusion has a spectrum of the Kolmogorov type and is fully developed, in the sense that the total turbulent energy density is roughly equal to the average energy density of the magnetic field (i.e., $\langle \delta B^2/B^2 \rangle \approx 1$). Most of the energy in this turbulent field resides at long wavelength, so it is clearly important to determine the maximum wavelength up to which the spectrum extends. Biermann and Strittmatter identify this length with the gyroradius of the most energetic hadrons, found by assuming saturation by either proton synchrotron losses or photo-pion production (Eq. 8). Assuming the Kolmogorov power law spectrum continues down to the wavelengths of importance for electron acceleration (which are a factor of 10^{10} shorter) one *predicts* that the synchrotron cut-offs should occur at about the frequency at which they are in fact observed. The main difficulty with this model is the question of whether or not the assumption about the turbulence spectrum is realistic: a typical eddy turn-over time, which is a reasonable estimate of the minimum time required for the development of a turbulence spectrum, is given by the ratio of the wavelength λ to the Alfvén speed, and is much longer (for a mildly relativistic shock) than the time available in the precursor, given approximately by $\lambda c/u_-^2$.

Roland, Pelletier and Muxlow, on the other hand, assume a $k^{-3/2}$ spectrum of turbulence, and identify the maximum permitted wavelength with the physical size of the accelerating region. Then, noting that observations place a limit on the position of the synchrotron cut-off (for the special case of Cyg A) they *deduce* that the level of turbulence is relatively low ($\langle \delta B^2/B^2 \rangle \approx 10^{-4}$).

Interestingly, each of these models predicts that hadrons are accelerated to extremely high energy at the hot spots in jets. Biermann and Strittmatter estimate 10^{12} GeV as the maximum energy of protons in such sources, which is consistent with Hillas' (1984) identification of the hot spots in the lobes of powerful radio galaxies as one of the few possible acceleration sites of the highest energy cosmic rays.

4 Efficiency

One way of estimating the efficiency of the first order Fermi process at a shock front is to look at the stationary solutions available to the system. For nonrelativistic shocks, this problem has been investigated in some detail (Drury and Völk 1981, Axford, Leer and McKenzie 1982, Achterberg, Blandford and Periwal 1984, Heavens 1984) using the so-called 'two-fluid' approximation. These papers were important in preparing the way for subsequent calculations of the full time dependent and momentum dependent problem (Bell 1987, Falle and Giddings 1987, Duffy 1990). In the relativistic case, progress has been understandably slower, and it is only recently that the basic two-fluid, stationary solutions have been elucidated (Baring and Kirk 1991).

The main types of solution found are the same as in the nonrelativistic analyses: a hydrodynamic shock in which the nonthermal particles (CRs) are a small perturbation, a CR dominated shock (which may however contain a gas subshock) and an intermediate structure which always has both a gas sub-shock and a precursor in which the CRs decelerate and compress the incoming plasma. If attention is restricted to CRs which are injected (by some unspecified process) at the gas subshock, then the first type of solution has no accelerated particles at all and can be discarded. For a given set of upstream conditions, there are then either none, one or two possible stationary solutions. This structure is illustrated in Fig. 1, where the efficiency, defined as the energy flux in CRs divided by the total energy flux, is plotted as a function of the Mach number of the flow far upstream, for various values of the upstream velocity. For the parameters chosen (effective ratio of specific heats of the gas and CRs of 8/5 and 4/3, respectively) there exists no stationary solution if the Mach number is less than a critical value which depends on the upstream speed. At low speed, the critical Mach number is smallest, so that there are no solutions at all for Mach numbers less than about 5. At Mach numbers higher than the critical value, two stationary solutions appear, one with high efficiency and one with low efficiency. Increasing the Mach number makes the low efficiency solution even less efficient, and can even cause it to reach zero efficiency and disappear if the upstream speed is sufficiently small (in this case

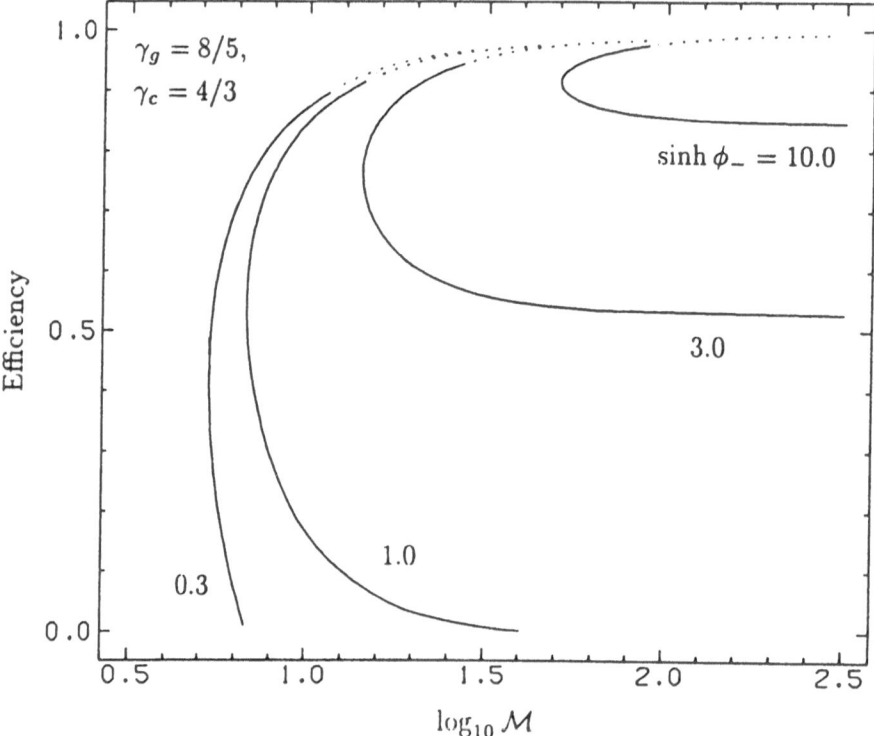

Fig. 1. the efficiency of particle acceleration by the first order Fermi process operating at a plane parallel shock front as a function of the Mach number M of the upstream flow for various values of the spatial component of the four velocity $\sinh\phi_-$ of the upstream flow expressed in units of the speed of light. The efficiency is defined as the ratio of the energy flux carried by the accelerated particles downstream to the total energy flux, measured in the shock frame. It is assumed that no accelerated particles are present far upstream. Solid lines depict stationary solutions which contain a precursor and a gas sub-shock. Dotted lines are solutions in which no sub-shock is present. The ratio of specific heats of the gas γ_g is assumed to be 8/5, that of the accelerated particles 4/3.

if the spatial component of the four-velocity is less than $5\sqrt{6}/12$ in units of c). The high efficiency (CR dominated) solution, on the other hand, becomes more and more efficient, until the cosmic rays completely smooth out the gas sub-shock. Although solutions continue, and are shown in Fig. 1 by the dotted line, it is not clear whether these structures are capable of injecting particles into the acceleration process. For low speed shocks, however, they are the only high Mach number stationary solutions, so that very strong nonrelativistic shocks may be expected quickly to become CR dominated, thus possibly switching off the injection process as the sub-shock is smoothed out. On the other hand, this need not happen to high speed (relativistic) shocks, since there exists a stationary solution of quite high efficiency (over 50% for a four-velocity of $3c$).

The stationary solutions discussed above are derived on the assumption that loss processes limiting the highest energy have no effect on the overall dynamics. This can be true only if the spectrum of accelerated particles is steeper than p^{-4}, since otherwise the particles of highest energy dominate the CR pressure. The actual spectrum produced by a particular shock depends, however, on the diffusion coefficient governing the particle transport. If this rises towards high momentum, then it may well be the case that loss processes intervene and permit new types of stationary structure. These have been discussed by Eichler (1984) and Ellison and Eichler (1984, 1985) for the case of nonrelativistic flows. In the relativistic case, it seems that Monte-Carlo simulations provide a promising approach to the problem (Ellison, Jones and Reynolds 1990).

References

Achterberg, A., Blandford, R.D. and Periwal, V. 1984 *Astron. Astrophys.* **132**, 97.

Axford, W.I., Leer, E. and McKenzie, J.F. 1982 *Astron. Astrophys.* **111**, 317.

Baring, M.G. and Kirk, J.G. 1991 *Astron. Astrophys.* in press.

Begelman, M.C. and Kirk, J.G. 1990 *Astrophys. J.* **353**, 66.

Bell, A.R. 1987 *Mon. Not. R. astr. Soc.* **225**, 615.

Biermann, P.L. and Strittmatter, P.A. 1987 *Astrophys. J.* **322**, 643.

Blandford, R.D and Eichler, D. 1987 *Phys. Rep.* **154**, 1.

de Hoffmann, F. and Teller, E. 1950 *Phys. Rev.* **80**, 692.

Drury, L.O'C. and Völk, H.J. 1981 *Astrophys. J.* **248**, 344.

Drury, L.O'C. 1983 *Reports Prog. Phys.* **46**, 973.

Duffy, P. 1990 Ph.D. thesis, Trinity College Dublin.

Eichler, D. 1984 *Astrophys. J.* **277**, 429.

Ellison, D.C. and Eichler, D. 1984 *Astrophys. J.* **286**, 691.

Ellison, D.C. and Eichler, D. 1985 *Phys. Rev. Letters* **55**, 2735.

Ellison, D.C., Jones, F.C. and Reynolds, S.P. 1990 *Astrophys. J.* **360**, 702.

Falle, S.A.E.G. and Giddings, J.R. 1987 *Mon. Not. R. astr. Soc.* **225**, 399.

Forman, M.A. and Webb, G.M. 1985 in *Collisionless Shocks in the Heliosphere: a Tutorial Review* eds: R.G. Stone and B.T. Tsurutani, Geophysical Monograph Series, **43**, 91.

Heavens, A.F. 1984 *Mon. Not. R. astr. Soc.* **210**, 813.

Hillas, A.M. 1984 *Ann. Rev. Astron. Astrophys.* **22**, 425.

Jokipii, J.R. 1982 *Astrophys. J.* **255**, 716.

Jokipii, J.R. 1988 in *Proceedings of the Sixth International Solar Wind Conference* eds: V.J. Pizzo, T.E. Holzer and D.G. Sime, NCAR Technical Note, **2**, 481.

Kirk, J.G. and Heavens, A.F. 1989 *Mon. Not. R. astr. Soc.* **239**, 995.

Kirk, J.G. and Schneider, P. 1987 *Astrophys. J.* **315**, 425.

Laing, R.A. 1988 *Nature* **331**, 149.

Ostrowski, M. 1991 to appear in *Mon. Not. R. astr. Soc.* , (see also this volume).

Quenby, J.J. and Lieu, R. 1989 *Nature* **342**, 654.

Roland, J., Pelletier, G. and Muxlow, T.W.B. 1988 *Astron. Astrophys.* **207**, 16.

Röser, H.-J. 1989, in *Hot Spots in Extragalactic Radio Sources*, ed. K. Meisenheimer and H.-J. Röser (Berlin: Springer) p. 91.

Sikora, M., Kirk, J.G., Begelman, M.C. and Schneider, P. 1987 *Astrophys. J. Letts.* **320**, L81.

Webb, G.M., Axford, W.I. and Terasawa T. 1983 *Astrophys. J.* **270**, 537.

Fermi Particle Acceleration in Relativistic Shocks:
Preliminary Nonlinear Results

Donald C. Ellison

Department of Physics, Code 8202, North Carolina State University,
Raleigh, NC 27695-8202

Abstract: We have extended a previously developed and tested Monte Carlo simulation of parallel shocks to include relativistic flow velocities. This steady-state simulation treats thermal particle injection, determines the shock structure in the presence of accelerated particles, and simultaneously and self-consistently calculates the ratio of specific heats in the shocked plasma and the effects of particle escape. We present preliminary results showing the absolute acceleration efficiencies and spectra for a range of parameters and show that relativistic shocks can be extremely efficient particle accelerators.

1 Introduction

Observational, theoretical, and computer simulation work has demonstrated that collisionless shocks can accelerate ambient ions to many times thermal energies. Non-thermal particle distributions have been directly observed by spacecraft at the quasi-parallel Earth bow shock and at quasi-parallel interplanetary traveling shocks and these distributions have been modeled successfully with the first-order Fermi particle acceleration mechanism (see Blandford and Eichler 1987). Less direct yet still convincing evidence suggests that relativistic electrons are shock accelerated in supernova remnant blast waves (see Reynolds 1988 for a review). These blast waves have long been suggested as sources of most galactic cosmic rays (e.g., Axford 1981), and the Fermi mechanism (Krymsky 1977; Axford et al. 1977; Bell 1978; Blandford and Ostriker 1978) is likely to produce these electrons and energetic cosmic ray ions as well. The ubiquitous nature of particle acceleration at astrophysical shocks is explained, in part, by recent large-scale plasma simulation results (e.g., Leroy and Winske 1983; Quest 1988; Burgess 1989) which show that particle acceleration occurs in both quasi-perpendicular and quasi-parallel shocks as an essential part of the dissipation process. While these plasma simulation results are currently restricted to small energy increases, they clearly show that particle acceleration is an intrinsic part of collisionless shocks.

The success of the first-order Fermi mechanism in describing particle acceleration in shocks with nonrelativistic flow velocities has prompted work on the more
difficult problem presented when flows become relativistic. Relativistic mass flows
are expected to occur in several important environments such as pulsar winds,
extragalactic radio jets, and possibly accretion flows onto compact objects such as
neutron stars or black holes.

Initial test-particle results of first-order Fermi acceleration in relativistic shocks
show quantitative differences from nonrelativistic shocks (e.g., Peacock 1981; Kirk
and Schneider 1987a,b; Ellison *et al.* 1990a). Perhaps the most important difference
is that relativistic shocks appear to be even more efficient particle accelerators
and able to produce flatter spectra than are nonrelativistic shocks. However, if
relativistic shocks are efficient accelerators, nonlinear feedback effects between the
accelerated particles and the shock become important and test-particle results may
no longer be a good approximation to the actual situation. These effects include the
production of magnetic turbulence via wave-particle interactions (e.g., Lee 1982)
and the large-scale slowing and heating of the unshocked plasma by backstreaming
accelerated particles (e.g., Drury and Völk 1981).

In this paper, we apply Monte Carlo techniques to calculate the nonlinear
shock structure (i.e., the smoothing of the plasma flow velocity produced by the
accelerated particles) under steady-state conditions. We do not consider the other
nonlinear process wherein the particle scattering is determined self-consistently
from the particle distribution function. Instead, we simply assume that particles
scatter with a mean free path, $\lambda \propto R^{\alpha}$, where $R = pc/(Ze)$ is the particle rigidity, p
is the momentum, Z is the charge number, α is a constant parameter, and c and e
have their usual meanings. This scattering law, with $\alpha = 1$, has been shown to be an
excellent approximation in low velocity heliospheric shocks (Ellison *et al.* 1990b),
but must be considered essentially *ad hoc* for relativistic plasmas.

The Monte Carlo simulation is completely general as far as the flow velocity is
concerned and the techniques apply equally well to relativistic and nonrelativistic flows. This offers a unique advantage because the simulation has been tested
directly against observations in nonrelativistic situations where spacecraft data is
available (e.g., Ellison and Möbius 1987). In addition, the code has produced test-
particle results for relativistic shocks (Ellison *et al.* 1990a) which are in excellent
agreement with the analytic results of Peacock (1981) and Kirk and Schneider
(1987b).

We find that nonlinear shock smoothing in relativistic shocks produces changes
in the spectral shape and overall acceleration efficiency much in the same way as
in nonrelativistic shocks (e.g., Ellison and Eichler 1984). If the diffusion coefficient
increases with particle energy (as is expected in all realistic cases), the resulting
particle spectrum will be concave, becoming flatter at higher energies, and the
shock smoothing will reduce the efficiency for accelerating low-energy ions and
regulate the number of ions accelerated to high energy (e.g., Eichler 1979).

Also, when the diffusion coefficient increases with energy, particles will escape
from high Mach number, finite-sized, steady-state shocks and carry away an appreciable fraction of the energy flux, Q_{esc}, before being convected far downstream.

Besides cutting off the spectrum at high energies, particle escape influences the overall acceleration efficiency. The energy loss, in effect, makes the shocked plasma more compressible and the more energy lost, the larger the shock compression ratio, r, becomes. This effect is strongly nonlinear because the acceleration efficiency increases with the shock compression ratio and energy loss must be included *explicitly* in the Rankine-Hugoniot jump conditions (e.g., Ellison and Eichler 1984).

A third effect on the compression ratio and acceleration efficiency occurs when relativistic particles contribute significantly to the downstream pressure. If this is the case (and it always will be in relativistic shocks), the effective ratio of specific heats, γ_{eff}, is reduced and approaches $4/3$. Since the resultant spectrum is strongly dependent on γ_{eff} and vice versa, γ_{eff} must be determined self-consistently by the calculation.

All of these nonlinear effects are included in the Monte Carlo simulation and are present in nonrelativistic shocks as well. While no qualitative differences seem to occur for relativistic shocks versus nonrelativistic shocks, we do find in these preliminary calculations that the particle spectra generated by relativistic shocks are *less* sensitive to changes in compression ratio than might be expected.

2 Monte Carlo Method

The simulation we have developed for a plane, steady-state, high Mach number, parallel shock has been described previously (e.g., Ellison *et al.* 1981; Ellison *et al.* 1990b). Briefly, we have used Monte Carlo techniques to iteratively obtain a shock flow velocity profile which conserves mass, momentum, and energy fluxes when ions are heated and accelerated in the compressive shock flow. We assume that ions of all energies make *isotropic and elastic scatterings* in the local plasma frame with a mean free path, λ. While the scattering is not determined self-consistently from the particle distribution function and the background magnetic field, evidence from plasma simulations (e.g., Quest 1988) and comparisons with bow shock observations at low energies (e.g., Ellison *et al.* 1990b) support the simple rigidity dependence and scattering scheme we assume in nonrelativistic flows.

It is also assumed that shock heated downstream ions can freely scatter back across the subshock into the upstream region without being thermalized. Such "thermal leakage" of downstream ions provides an injection mechanism for Fermi acceleration since a few ions will manage to scatter across the subshock several times and receive repeated energy gains. The model treats thermal particle injection and acceleration as an inherent part of the shock dissipation mechanism and unifies parallel shock structure with first-order Fermi acceleration. The slowing and heating of the unshocked plasma, mandated by the presence of accelerated ions ahead of the shock, strongly influences thermal ion injection: the more the unshocked plasma is slowed, the smoother the shock and the fewer injected ions. Only when particle injection and acceleration are coupled through the pressure of the accelerated particles on the unshocked plasma can quantitative predictions for the absolute injection and acceleration efficiencies and complete ion distribution functions over all energy ranges be made.

The model also includes the escape of energetic particles at a high-energy cutoff, E_{max}. When particles obtain an energy greater than E_{max}, they are assumed to freely leave the shock system. As mentioned above, particle escape from finite size, steady-state shocks will result in an increase in the compression ratio and an increase in the acceleration efficiency.

The mean free path is taken to be $\lambda \propto R^\alpha/\rho = \lambda_0(A/Z)^\alpha(\gamma v/u_2)^\alpha[\rho_2/\rho(x)]$, where ρ is the plasma density, u_2 is the downstream flow velocity, A/Z is the ratio of mass number to charge number, $\gamma = 1/\sqrt{[1-(v^2/c^2)]}$, λ_0 is the downstream mean free path of a particle with $A/Z = 1$ and $v = u_2/\gamma$, and the particle momentum and velocity are measured in the local plasma frame.[1] As mentioned above, $\alpha = 1$ produces excellent fits to spacecraft observations of accelerated particles at the Earth's bow shock and we continue to use $\alpha = 1$ in what follows.

It has been shown (Kirk and Schneider 1987a) that the spectrum produced in relativistic shocks depends on the type of scattering assumed. In particular, large-angle scattering (where the direction the particle moves is randomized at each scattering) produces flatter spectra than pitch-angle diffusion. This effect comes about because the distribution function in the local plasma frame is not nearly isotropic in the shock frame and the average angle a particle's velocity vector makes upon crossing the shock will depend on its previous history. In nonrelativistic shocks, when $v \gg u$, the distribution function is nearly isotropic in all reference frames and the power law index of the accelerated spectrum is independent of the scattering. In this work, large-angle scattering is assumed throughout.

3 Results

In the top panel of Figure 1 we show the shock structure (i.e., the flow velocity as a function of position) for a shock with velocity $u_1 = 0.95c$, far upstream temperature $T_1 = 1 \times 10^8$ K, and a maximum cutoff kinetic energy of $E_{max} = 1 \times 10^8$ keV. The smoothed shock profile (solid line) was determined in an iterative fashion in the following way. Both the shape of the profile and the overall compression ratio must be iterated simultaneously. We start with a discontinuous or test-particle shock structure (dashed line in top panel), and an initial r ($r \equiv u_1/u_2$). Particles are injected far upstream and are scattered and convected through the shock and are accelerated if they scatter across the sharp transition. After some number of particles have been run through the shock, the momentum and energy fluxes are calculated at various positions along the x-axis. These fluxes are shown in the lower panel of Figure 1. The upper dashed and dotted lines in the lower panel are the momentum and energy fluxes, respectively, calculated with the discontinuous shock profile of the upper panel. The discontinuous shock doesn't conserve momentum or energy and the downstream fluxes are well above the far upstream values.

---

[1] The subscript 1 implies far upstream values and 2 implies far downstream values.

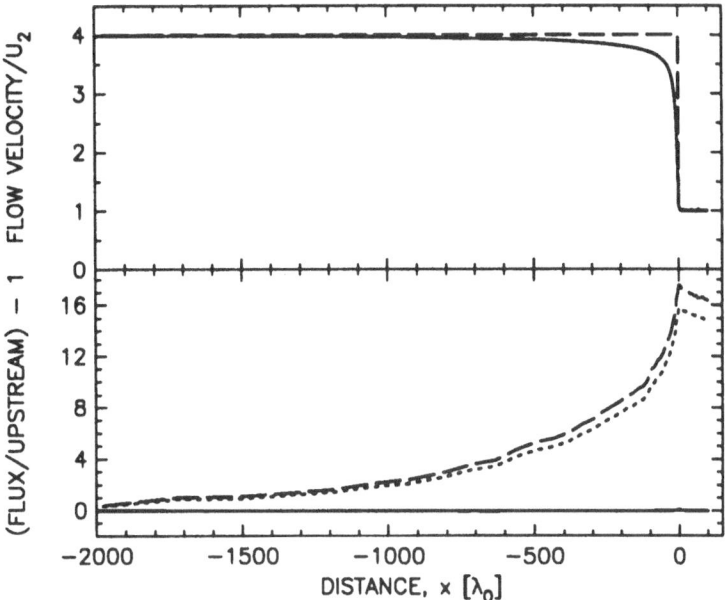

Fig. 1. Shock profiles (upper panel) and momentum and energy fluxes (lower panel).

For a given r, the code iterates the shock profile, attempting to find one that will conserve the momentum and energy fluxes across the shock. The simulation is run again with this new profile and particles are again injected far upstream and propagated through the shock. The only difference now is that, due to the smooth shock structure, particles can gain (or lose) energy on every scattering. However, since the conservation of flux will depend on the value of r, if the "correct" r is not used, the fluxes will not be constant across the shock. The compression ratio is then varied and the shock profile iterated again until the fluxes are as constant as possible. We find that a unique r and profile exist which allows simultaneous conservation of momentum and energy fluxes. Figure 1 shows the final momentum and energy fluxes produced by the smooth profile with $r = 4.0 \pm 0.3$ (note that there are two lower curves in the lower panel). The convergence is quite good with no deviations from the far upstream value greater than $\pm 10\%$. The number flux is trivially conserved in all cases.

One important aspect of the smooth profile in the top panel of Figure 1 is the presence of a "subshock" with a length-scale on the order of an upstream thermal particle convection length. This subshock heats the incoming cold gas and produces a hot downstream particle population which provides seed particles for further acceleration. Some fraction of the hot downstream plasma recrosses the shock and becomes Fermi accelerated. We find, in contrast to analytic two-fluid results discussed below, that a subshock exists regardless of the acceleration efficiency as long as only thermal ions are injected and accelerated by the shock.

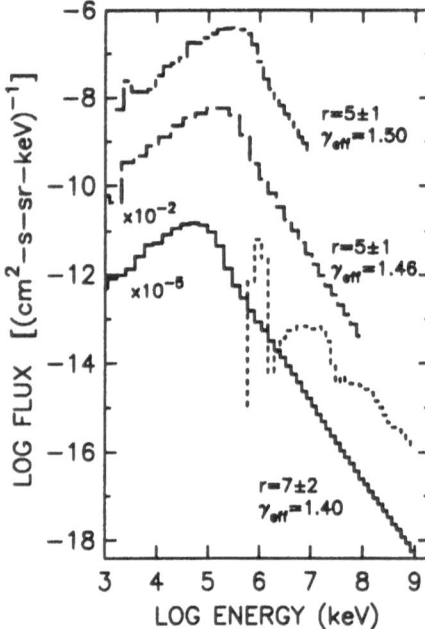

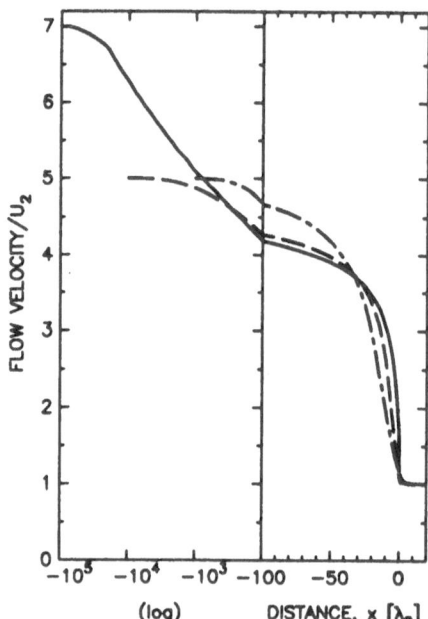

Fig. 2. Differential energy spectra calculated in the shock frame. The numbers on the left indicate the factors curves have been displaced for clarity.

Fig. 3. Shock profiles producing spectra shown in Figure 2. Note the divided distance scale.

In Figure 2 we show downstream energy spectra which were obtained with a constant shock velocity and far upstream temperature, but with changing E_{max}. The dot-dashed, dashed, and solid curves have $E_{max} = 10^7$, 10^8, 10^9 keV, respectively, all for $u_1 = 0.9c$ and $T_1 = 1 \times 10^8$ K. In each case the shock profile and compression ratio were determined iteratively as just described. As E_{max} increases, several important effects occur: (1) the overall shock compression ratio increases from $r \simeq 5$ to 7; (2) γ_{eff} decreases from ~ 1.5 to 1.4; (3) the fraction of incoming energy flux which escapes at E_{max} increases from $Q_{esc} \simeq 0.2$ to 0.3; (4) the thermal peak shifts to lower energy; and (5) the concave spectral shape, while still slight, becomes more noticeable.

These effects are, of course, related and the increase in r comes about because the spectra with larger E_{max} have a greater fraction of the particle pressure in relativistic particles so the shocked γ_{eff} is closer to 4/3, and also because a greater fraction of energy flux is lost at the high energy cutoff. Ellison and Eichler (1984) describe these effects for nonrelativistic shocks and we find similar behavior here for relativistic flows.

The dotted curve in Figure 2 shows the *test-particle* spectrum produced for a shock with the same parameters as the $E_{max} = 1 \times 10^9$ keV example (i.e., same u_1, T_1, and r). The test particle spectrum shows humps which result from individual shock crossings. When the shock is allowed to smooth, these humps disappear (solid curve), the spectrum becomes steeper, and the energy is distributed more evenly between the thermal peak and the highest energy particles. It is clear that the test-particle result doesn't even approximate the self-consistent calculation, showing how important nonlinear effects are. In particular, the self-consistent spectrum doesn't flatten to the test-particle slope at the highest energies. The spectral index for the solid curve (calculated between 1×10^7 and 1×10^9 keV) is -1.7 versus -1.35 for the test-particle spectrum.

In Figure 3 we show the shock profiles which produced the spectra shown in Figure 2. The long-scale smoothing of the shock, which is proportional to the upstream diffusion length, $\kappa / <u>$, where $<u>$ is some average value of the upstream flow velocity, increases with increasing $E_{\max}$. However, even though the overall shock becomes very broad, a distinct subshock exists in all cases.

We define the acceleration efficiency, $\epsilon(> E)$, at or behind the shock, to be the fraction of total energy flux contained in particles with energy above E (including the energy flux that escapes at $E_{\max}$). This is shown in Figure 4 for the three examples given in Figure 2. As $E_{\max}$ increases, a greater fraction of the energy flux is contained in the highest energy particles. This is a well-known property of high Mach number, steady-state shocks with energy dependent diffusion coefficients (e.g., Eichler 1984; Ellison and Eichler 1984); the shock becomes more efficient as $E_{\max}$ increases and a greater fraction of the energy flux is lost at $E_{\max}$.

The shock with $E_{\max} = 1 \times 10^9$ keV puts almost 90% of the total energy flux into particles with energies above 1 GeV, and $\sim 60\%$ into particles with energies above 100 GeV.

In Figure 5 we show the partial pressure, i.e., $(j/3)\rho_E$, where ρ_E is the energy density per unit energy and $j = (E + 2m_oc^2)/(E + m_oc^2)$ is equal to 2 for non-relativistic particles and 1 for relativistic particles. Figure 5 shows that as $E_{\max}$ increases, the particle pressure becomes divided between the thermal peak, which shifts to lower energy, and the highest energy particles with a smaller and smaller fraction at intermediate energies. The concave nature of the spectra in Figure 2 becomes clearly evident in the bowl shape of the partial pressure. The partial pressure does not include the escaping energy flux.

4 Conclusions

We have performed a steady-state calculation of relativistic parallel shocks which combines first-order Fermi particle acceleration with a determination of the shock structure. Particles are accelerated directly from the thermal plasma and the shock structure is determined such that mass, momentum, and energy fluxes, including the contribution from accelerated particles, are conserved at all positions relative to the shock. A shock profile with essentially two length scales is produced; a long-scale smooth shock produced by energetic particles, and an imbedded subshock which heats the thermal plasma. The shock smoothing lowers the Mach number of the subshock and reduces and regulates the acceleration efficiency of the high energy particles.

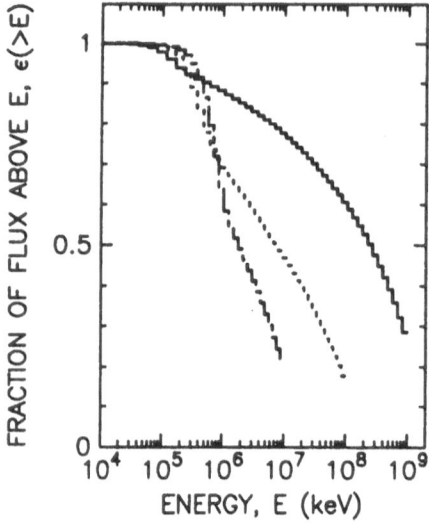

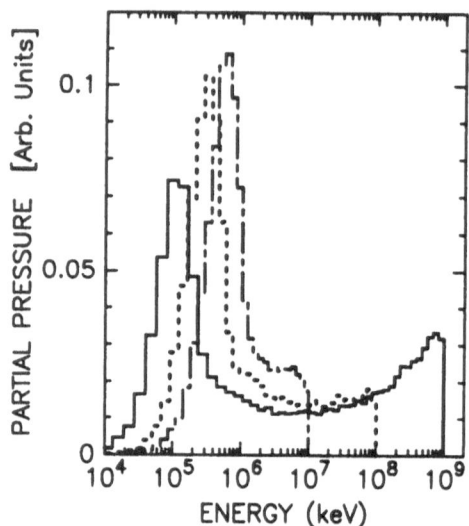

Fig. 4. Fraction of energy flux above E versus kinetic energy, E. The endpoints of each curve indicates the fraction of energy flux lost at E_{max}.

Fig. 5. Partial pressure versus energy for the three examples shown in Figure 2. The dot-dashed curve shows the $E_{max} = 1 \times 10^7$ keV example, the dotted curve shows the $E_{max} = 1 \times 10^8$ keV example, and the solid line shows the $E_{max} = 1 \times 10^9$ keV example.

Our calculation includes the self-consistent determination of the shock compression ratio, r, and the ratio of specific heats, γ_{eff}, for any combination of relativistic and nonrelativistic particles. Since both of these quantities strongly influence (and depend on) the acceleration efficiency, their determination must be an intrinsic part of the calculation of particle acceleration, they cannot be treated as parameters. In addition, we include energy escape at a high energy cutoff. In all realistic steady-state shock acceleration problems, particles will escape from high Mach number shocks and carry away a dynamically significant fraction of the energy flux. As with r and γ_{eff}, this energy loss both influences and depends on the acceleration efficiency and must be included self-consistently in the overall shock structure-acceleration problem.

We find that relativistic shocks are, indeed, efficient accelerators and can easily put over 50% of the energy flux into very energetic particles (see Figure 4). These shocks can also produce flat spectra, but the spectral shape is strongly modified from that predicted by test-particle calculations and does not obtain the test-particle limit at the highest energies, the modified spectra remain considerably steeper.

Monte Carlo techniques are well suited to the problems associated with relativistic shocks and require little modification from previous nonrelativistic applications. While there are no important qualitative differences from solutions of nonrelativistic shocks which include a high energy cutoff and self-consistent injection from the thermal plasma (e.g., Ellison and Eichler 1984), the large spatial scales make finding a solution for relativistic shocks more difficult. For this reason our preliminary solutions span a limited range of parameters and show fairly large error bars.

Of particular importance are the strong differences between our results and those from hydrodynamic, two-fluid shock solutions. The two-fluid approach, first presented by Drury and Völk (1981), is an analytic technique, based on the diffusion equation, which treats the thermal gas and energetic particles as separate fluids connected only by a parameterized injection rate. The energetic particles are decoupled from the thermal plasma insofar as the jump across the thermal shock is concerned. For a more complete look at such solutions one should refer to Drury (1983) where it is pointed out that three basic types of solutions occur depending on the input parameters. In particular, for large Mach numbers, the downstream state is not always uniquely determined from the upstream state. For certain low values of the upstream cosmic ray pressure, P_{c1}, at least two values for the downstream pressure are possible. In fact, Drury and Völk (1981) predict that for Mach numbers above about 6, large downstream cosmic ray pressures can result even for a zero value of P_{c1}. The energetic particle pressure is created out of nothing! This can occur even where no thermal shock exists. The implication is that even with no thermal gas subshock to inject particles and no upstream energetic seed particles, these shocks could still generate cosmic rays.

Baring and Kirk (1990) did a relativistic two-fluid calculation and restricted their treatment to profiles which included subshocks, though they made no self-consistent inclusion of particle injection at the discontinuity. They found that, within the confines of the assumptions of this model, only weak shocks can accelerate particles. This clearly contradicts a host of observational evidence.

Such unphysical solutions appear to us to arise out of the lack of conservation of energetic particle flux in the set of equations governing the shocks. Conservation of cosmic ray particle number must hold unless there is explicit injection, and such a constraint would prevent these "something from nothing" solutions.

Another feature of the steady-state two-fluid models that contributes to the existence of unphysical solutions is the lack of spectral information and with it the lack of a cutoff at high energy. This limits realistic solutions to low Mach numbers.

There have been several important extensions and generalizations of the two-fluid method including addition of interactions between the energetic particles and Alfvén waves (e.g., Völk et al. 1984), a self-consistent calculation of γ_{eff} and an energy dependent diffusion coefficient (Heavens 1983, 1984), and time-dependent calculations (Falle and Giddings 1987 and Bell 1987). However, all of the two-fluid solutions we are aware of, steady-state and time-dependent, show the subshock vanishing for high Mach numbers even in the limit when no upstream cosmic rays are injected.

An alternative analytic treatment of modified shocks which avoids many of the problems intrinsic to the two-fluid technique and allows the calculation of the particle spectrum was developed by Eichler (1979, 1984) (see also Krymsky

1981). Eichler is able to solve the diffusion equation including energy escape by taking advantage of approximations resulting from the assumed rapid increase in κ with energy. In contrast to two-fluid models, it is predicted that the acceleration efficiency increases monotonically with Mach number, and cosmic ray dominated shocks do not occur (note that pre-existing cosmic rays are not injected in Eichler's solution). As a consequence, there will always be some shock heating, an important consideration when interpreting observations of thermal X-rays in astrophysical objects. The results we present here are totally consistent with these analytic results (i.e., Ellison and Eichler 1984), indicating that no qualitative differences result in Fermi particle acceleration when relativistic flows occur.

References

Axford, W. I., 1981, in *Proc. 17th Int. Cosmic Ray Conf.* (Paris), **12**, 155.

Axford, W. I. , Lear, E. , and Skadron, G., 1977, in *Proc. 15th Int. Cosmic Ray Conf.* (Plovdiv), **11**, 132.

Baring, M. G., and Kirk, J. G., 1990, *Astr. Astrophy.*, in press.

Bell, A. R., 1978, *M.N.R.A.S.*, **182**, 147.

Bell, A. R., 1987, *M.N.R.A.S.*, **225**, 615.

Blandford, R. D. and Eichler, D., 1987, *Physics Reports* **154**, 1.

Blandford, R. D., and Ostriker, J. P., 1978, *Ap. J.(Letts)*, **221**, L29.

Burgess, D., 1989, *Geophys. Res. Lett.*, **16**, 163.

Drury, L. O'C., 1983, *Rep. Prog. Phys.* **46**, 973.

Drury, L. O'C. and Völk, H. J., 1981, *Ap. J.*, **248**, 344.

Eichler, D., 1979, *Ap. J.*, **229**, 419.

Eichler, D., 1984, *Ap. J.*, **277**, 429.

Ellison, D. C., and Eichler, D., 1984, *Ap. J.*, **286**, 691.

Ellison, D. C., Jones, F. C., and Eichler, D., 1981, *Journal of Geophysics*, **50**, 110.

Ellison, D. C., Jones, F. C., and Reynolds, S. P., 1990a, *Ap. J.*, **360**, 702.

Ellison, D. C., and Möbius, E., 1987, *Ap. J.*, **318**, 474.

Ellison, D. C., Möbius, E., and Paschmann, G., 1990b, *Ap. J.*, **352**, 376.

Falle, S. A. E. G., and Giddings, J. R., 1987, *M.N.R.A.S.*, **225**, 399.

Heavens, A. F., 1983, *M.N.R.A.S.*, **204**, 699.

Heavens, A. F., 1984, *M.N.R.A.S.*, **210**, 813.

Kirk, J. G., and Schneider, P., 1987a, *Ap. J.*, **315**, 425.

Kirk, J. G., and Schneider, P., 1987b, *Ap. J.*, **322**, 256.

Krymsky, G. F., 1977, *Dokl. Akad. Nauk SSSR*, **243**, 1306.

Krymsky, G. F., 1981, *Izv. Akad. Nauk SSSR, Ser. Fiz.*, **45**, 461 (in Russian).

Lee, M. A., 1982, *J. Geophys. Res.*, **87**, 5063.

Leroy, M. M., and Winske, D., 1983, *Ann. Geophysicae*, **1**, 527.

Peacock, J. A., 1981, *M.N.R.A.S.*, **196**, 135.

Quest, K. B., 1988, *J. Geophys. Res.*, **93**, 9649.

Reynolds, S. P., 1988, in *Galactic and Extragalactic Radio Astronomy*, eds. G. L. Verschuur and K. I. Kellermann, Springer-Verlag, New York, p. 439.

Völk, H. J., Drury, L. O'C., and McKenzie, J. F., 1984, *Astr. Astrophys.*, **130**, 19.

Cosmic Ray Transport in AGNs and Jets

G. M. Webb

Department of Planetary Sciences, Lunar and Planetary Laboratory,
University of Arizona, Tucson, AZ 85721, U.S.A.

Abstract: The transport of cosmic rays in astrophysical fluid flows thought to be present in AGN's and jets is discussed in terms of diffusive particle transport equations; pitch angle evolution transport equations; and hydrodynamical equations governing the interaction of the energetic particles with the background flow. As an example of the use of these equations a discussion is given of the transport of energetic charged particles in an accretion flow onto a Schwarzschild black hole.

1 Introduction

The observations of nonthermal radiation in extragalactic radio sources (Meisenheimer et al. 1989) suggest the presence of both relativistic particles, and in some cases relativistic flows. The nonthermal emission from extended extragalactic radio sources is commonly interpreted as synchrotron radiation of relativistic electrons in a magnetic field. The observed frequency dependence of the radio flux at frequency ν, $S(\nu) \propto \nu^{-\alpha}$ implies a power law energy distribution of energetic electrons with number density distribution $N(E) \propto E^{-2\alpha-1}$ at energy E, with α canonically lying in the range $0.5 < \alpha < 0.9$ (e.g. Bridle and Perley 1984).

Two mechanisms commonly invoked to accelerate energetic particles in radio jets and compact objects are stochastic second order Fermi acceleration due to turbulent motion in a magnetized plasma or low frequency plasma and MHD waves (e.g. Achterberg 1981; Melrose 1980), and shock acceleration (regular Fermi acceleration) at shocks in the flow (see e.g. the reviews by Drury 1983; Blandford and Eichler 1987; Berezhko and Krimsky 1988). Two further mechanisms, not so commonly invoked are particle acceleration by scattering back and forth across a shear flow (e.g. Berezhko and Krimsky 1981; Earl, Jokipii and Morfill 1988; Webb 1989, 1990; Ostrowski 1990) and particle acceleration in a rotating scattering medium (Webb and Jokipii 1990). Particle acceleration in shear flows may be a significant mechanism in radio jets in the fluid velocity shear between the jet axis and the outer cocoon of the jet (e.g. Webb 1990; Ostrowski 1990).

111

Transport equations for cosmic rays in plasma flows supporting a spectrum of waves or turbulence that scatter the particles have been derived for both non-relativistic flows (e.g. Parker 1965; Gleeson and Axford 1967; Dolginov and Top-tygin 1966; Skilling 1975; Earl, Jokipii and Morfill 1988) and also for relativistic flows (e.g. Webb 1985, 1989; Kirk, Schlickeiser and Schneider 1988; Krülls 1990).

In the present paper we discuss particle transport and acceleration via the use of diffusive particle transport equations (Section 2). We then go on to illustrate the use of pitch angle dependent transport equations and hydrodynamical cosmic ray transport equations to describe energetic particle transport in an accretion flow onto a Schwarzchild black hole (Section 3). Section 4 concludes with a summary and overview.

2 Diffusive Particle Transport Equations

The underlying equation governing particle transport in a scattering medium is the relativistic Boltzmann equation, including the effects of radiation reaction terms (see Hakim 1967; Krülls 1990). For diffusive particle transport, the mean distribution function $f_0(\mathbf{x}, p')$ averaged over all momentum directions satisfies the diffusive transport equation (Webb 1989)

$$
\nabla_\alpha [cu^\alpha f_0 + q^\alpha] + \frac{1}{p'^2} \frac{\partial}{\partial p'} \left[-\frac{(p')^3}{3} f_0 \nabla_\beta u^\beta - p'(p'^0)^2 \dot{u}_\alpha q^\alpha \right.
$$
$$
\left. - \Gamma p'^4 \tau \frac{\partial f_0}{\partial p'} - p'^2 D_{pp} \frac{\partial f_0}{\partial p'} + <\dot{p}>_\ell f_0 \right] = 0
$$
(2.1)

where

$$
q^\alpha = -K^{\alpha\beta} \left\{ \nabla_\beta f_0 - \left[\frac{(p'^0)^2}{p'} \right] \dot{u}_\beta \frac{\partial f_0}{\partial p'} \right\},
$$
(2.2)

is the diffusive particle flux, including the relativistic heat inertia term ($\propto \dot{u}_\alpha$); u^α is the fluid velocity four vector; and

$$
\dot{u}_\alpha = u^\theta \nabla_\theta u_\alpha,
$$
(2.3)

is the acceleration vector of the fluid. In equation (2.1) x^α denotes the position four vector of the particle; ∇_α denotes covariant space time derivatives; p' denotes the magnitude of the particle 3-momentum in the scattering frame Σ' whereas $E' = p'^0 c$ is the total particle energy in Σ'. $K^{\alpha\beta}$ is the diffusion tensor including diffusion parallel and perpendicular to the mean magnetic field, as well as antisymmetric terms representing the Hall current (or particle drifts).

The first two terms in equation (2.1) represent particle transport via convection and diffusion, with the remaining momentum derivative terms representing particle energy changes. The third term proportional to the fluid velocity four divergence $\nabla_\beta u^\beta$ represents particle energy changes due to adiabatic compressions or expansions of the fluid. This term plays a central role in the acceleration of energetic particles at shocks in the first order Fermi mechanism. The fourth term in (2.1), dependent on the acceleration vector of the fluid $\dot{u}_\alpha$ clearly represents non- inertial energy changes associated with the fact that the particle momentum p' is measured relative to a local Lorentz frame moving with the fluid. It may be thought of as a gravitational redshift term, and has the form $F_\alpha q^\alpha$ ($F_\alpha = -p'(p'^0)^2 \dot{u}_\alpha$) which is reminiscent of the form of particle energization by a force field F_α. The remaining terms in equation (2.1) represent particle energization by shear (i.e. cosmic ray viscosity); second order Fermi acceleration and particle energy losses respectively (e.g. synchrotron losses; inverse Compton losses; nuclear collisions, etc). Strictly speaking, the fluid velocity four vector u^α in equation (2.1) should be replaced by the four velocity w^α of the mean scattering frame, but the distinction between these two velocities is negligible for small enough wave speeds (e.g. the Alfvén speed) of the scattering wave field relative to the background flow.

The viscous energization coefficient Γ depends in general on both the shear tensor $\sigma_{\alpha\beta}$ of the fluid:

$$\sigma_{\alpha\beta} = \nabla_\beta u_\alpha + \nabla_\alpha u_\beta + u_\beta \dot{u}_\alpha + u_\alpha \dot{u}_\beta - \frac{2}{3}(g_{\alpha\beta} + u_\alpha u_\beta)\nabla_\theta u^\theta, \qquad (2.4)$$

the magnetic field geometry; and whether the particle scattering is strong ($\omega\tau \ll 1$) or weak ($\omega\tau \gg 1$). Here $\omega = ZeB'/m'$ is the particle gyro- frequency, τ is the collision time, and $g_{\alpha\beta}$ is the metric tensor. For the case of strong scattering Γ depends only on the shear tensor of the fluid and is given by

$$\Gamma = \frac{c^2}{30}\sigma_{\alpha\beta}\sigma^{\alpha\beta}. \qquad (2.5)$$

For the opposite limit of weak scattering ($\omega\tau \gg 1$) $\Gamma \simeq (c^2/20)(\sigma'_{11})^2$, where the 1-direction lies along the magnetic field (see Webb 1989).

In the absence of extraneous energy losses ($<\dot{p}>_\ell = 0$), and second order Fermi acceleration ($D_{pp} = 0$), and for a rigidly rotating flow (in which $\nabla_\alpha u^\alpha = 0, \sigma_{\alpha\beta} = 0$), particles are accelerated solely by their interaction with the accelerating flow (i.e. the $\dot{u}_\alpha$ term in 2.1). In this idealized limit, the particles on average are forced to corotate with the flow and are energized in a fashion similar to that of a bead flung outward on a frictionless rigidly rotating wire (Webb and Jokipii 1990; McKenzie, Ip and Axford 1979).

Despite its apparent generality the diffusive particle transport equation is a poor approximation, when the particle anisotropies in momentum space are large or if the particle mean free path λ becomes comparable or greater than the scale lengths L_f and L_u for the variation of f_0 and u. A notable example in which this occurs is for particle acceleration by the first order Fermi mechanism at ultra-relativistic shocks (e.g. Kirk and Schneider 1987) or at non-relativistic shocks when the particle and fluid speeds are comparable (e.g. Kirk 1988). For the case of parallel shocks, it suffices to solve the evolution equation for the pitch angle distribution function $f(\mathbf{x}, p', \mu')$ where $\mu' = \mathbf{e}'_p \cdot \mathbf{B}'/B'$ is the pitch angle cosine and p' is the particle momentum measured in the scattering frame. The detailed form of this equation may be found for example in Webb (1985); Kirk, Schlickeiser and Schneider (1988) or Krülls (1990) (the coefficient of the ∇F term of equation (8.5) of Webb (1985) should be multiplied by the Lorentz gamma of the flow).

3 Particle Transport in an Accretion Flow Onto a Schwarzchild Black Hole

As an example of a flow in which both the evolution of the particle pitch angle distribution and general relativistic effects play a role consider the case of particle transport in a radial accretion flow onto a Schwarzchild black hole, with space-time metric

$$ds^2 = -d\tau^2 = -[-A^2 c^2 dt^2 + A^{-2} dr^2 + r^2 d\theta^2 + r^2 \sin^2\theta d\phi^2], \tag{3.1}$$

where

$$A = [1 - 2GM/(c^2 r)]^{1/2}, \tag{3.2}$$

G is the universal gravitating constant, M is the central gravitating mass and (r, θ, ϕ) are the usual spherical polar coordinates.

Assume that the particle distribution function f measured in the local tetrad frame Σ' moving with the radial inflow is symmetric (in momentum space) about the radial direction and independent of θ and ϕ (i.e. $f = f(r, t, p', \mu')$ where $\mu' = \cos\Theta'$, with Θ' the angle between the particle momentum and radial direction in Σ'). For the case of a radial, (or non-existent) mean magnetic field, and assuming pitch angle scattering of the particles, the energetic particle transport equation is

$$\frac{\gamma}{A}(p'^0 + p'\mu'\beta)\frac{\partial f}{\partial x^0} + \gamma A(\beta p'^0 + p'\mu')\frac{\partial f}{\partial r} + G^p \frac{\partial f}{\partial p'} + G^\mu \frac{\partial f}{\partial \mu'}$$
$$= \frac{p'^0}{c}\frac{\partial}{\partial \mu'}\left[\frac{1 - \mu'^2}{2}\nu\frac{\partial f}{\partial \mu'}\right], \tag{3.3}$$

where

$$G^p = -p'^0 p'\left\{\frac{A\gamma\beta}{r}(1 - \mu'^2) + \mu'^2\left[\frac{\partial}{\partial r}(\gamma\beta A) + \frac{\partial}{\partial x^0}\left(\frac{\gamma}{A}\right)\right]\right\}$$
$$- (p'^0)^2 \mu'\left[\frac{\partial}{\partial x^0}\left(\frac{\gamma\beta}{A}\right) + \frac{\partial}{\partial r}(A\gamma)\right], \tag{3.4}$$

$$G^\mu = -\frac{(1 - \mu'^2)}{p'}\left\{(p'^0)^2\left[\frac{\partial}{\partial x^0}\left(\frac{\gamma\beta}{A}\right) + \frac{\partial}{\partial r}(A\gamma)\right] - p'^2\frac{A\gamma}{r}\right.$$
$$\left. + p'p'^0\mu'\left[\frac{\partial}{\partial r}(\gamma\beta A) + \frac{\partial}{\partial x^0}\left(\frac{\gamma}{A}\right) - \frac{A\gamma\beta}{r}\right]\right\}. \tag{3.5}$$

In equations (3.3)–(3.5) $x^0 = ct, \beta = V/c$ is the radial inflow speed; and $\gamma = (1 - \beta^2)^{-1/2}$ is the Lorentz gamma of the flow. Note that synchrotron and inverse Compton losses have not been included in equation (3.3) and we assume the particle scattering is via pitch angle scattering with pitch angle scattering frequency ν. Equations (3.3)–(3.5) follow directly from equations (8.1) and (8.2) of Webb (1985), by evaluating the comoving frame affine connection coefficients $\Gamma'^\alpha_{\beta\gamma}$. The derivation is essentially the same as that carried out by Riffert (1986) for the case of photon transport, in which an appropriate Lorentz boost is made between the comoving tetrad frame and a stationary local tetrad frame.

Corresponding hydrodynamical equations describing the energy and momentum transfer between the cosmic rays and background thermal gas are readily obtained by using the Eddington approximation (see Webb 1987, and Baring and Kirk 1990 for similar hydrodynamical models of cosmic ray modified shocks). Assuming the cosmic rays constitute a hot low density gas with negligible mass density, the hydrodynamical equations governing the interaction of the two fluids may be written as

$$\frac{\partial}{\partial x^0}\left(\frac{\gamma n}{A}\right) + \frac{1}{r^2}\frac{\partial}{\partial r}(r^2 A\beta\gamma n) = 0, \tag{3.6}$$

$$\nabla_\beta T_c^{\alpha\beta} = D^\alpha; \quad \nabla_\beta T_g^{\alpha\beta} = -D^\alpha, \tag{3.7}$$

where n is the thermal gas number density, $T_c^{\alpha\beta}$ and $T_g^{\alpha\beta}$ denote the stress energy tensors for the cosmic rays and thermal background gas and

$$D'^\alpha = [0, -\frac{\bar\nu}{c}T_c'^{10}, 0, 0] \tag{3.8}$$

represents the comoving frame momentum transfer between the cosmic rays and thermal gas. In the scattering frame Σ', the cosmic rays scatter magnetostatically without change of energy. Here $\bar\nu$ is an average of the pitch angle scattering frequency ν.

The background thermal gas is assumed to be a perfect gas with stress energy tensor

$$T_g^{\alpha\beta} = \frac{1}{c}[(E_g + m_0 nc^2)u^\alpha u^\beta + P_g(g^{\alpha\beta} + u^\alpha u^\beta)], \tag{3.9}$$

where $m_0 c^2$ is the mean rest mass energy of a gas particle and E_g and P_g are the internal energy density and pressure of the gas respectively. The corresponding comoving stress energy tensor for the cosmic-rays is of the form

$$T_c'^{\alpha\beta} = \frac{1}{c}\left\{ E_c\delta^{\alpha 0}\delta^{\beta 0} + P_{c\|}\delta^{\alpha 1}\delta^{\beta 1} + P_{c\perp}(\delta^{\alpha 2}\delta^{\beta 2} + \delta^{\alpha 3}\delta^{\beta 3}) \right.$$
$$\left. + cT_c'^{10}(\delta^{\alpha 1}\delta^{\beta 0} + \delta^{\alpha 0}\delta^{\beta 1}) \right\}. \tag{3.10}$$

The above equations are closed by assuming appropriate equations of state for the thermal gas and cosmic-rays. The simplest closure scheme is to assume the thermal and cosmic ray gases have adiabatic indices γ_g and γ_c, with

$$E_g = \frac{P_g}{(\gamma_g - 1)}, \quad E_c = \frac{P_{c\|} + 2P_{c\perp}}{3(\gamma_c - 1)} = \frac{P_{c\|}}{\gamma_c^* - 1}, \tag{3.11}$$

where

$$\gamma_c^* = 1 + 3e(\gamma_c - 1), \quad e = P_{c\|}/(P_{c\|} + 2P_{c\perp}), \tag{3.12}$$

defines the "effective" adiabatic index γ_c^* of the cosmic rays in terms of the Eddington factor e, which is assumed to be a known function of position r. However to obtain a fully self consistent treatment, the Eddington factor e must be determined by solving the pitch angle evolution equation (3.3) for f simultaneously with the overall hydrodynamical constraints (3.6) and (3.7) - a decidedly nontrivial, non linear problem.

For time independent, spherical flows the system of equations has two obvious integral, namely the number density conservation and energy conservation equations:

$$r^2 A \gamma \beta c n = J_n, \tag{3.13}$$

$$r^2 A^2 [\gamma^2 \beta c(W_g + W_c) + (2\gamma^2 - 1)c^2 T_c''^{10}] = J_E. \tag{3.14}$$

The comoving gas energy equation reduces to the first law of thermodynamics for the thermal gas:

$$\frac{dE_g}{dn} - \frac{(P_g + E_g)}{n} = 0 \quad \text{or} \quad P_g n^{-\gamma_g} = \text{const.}. \tag{3.15}$$

In these equations

$$W_g = E_g + P_g + m_0 n c^2, \quad W_c = E_c + P_{c\|}, \tag{3.16}$$

represent the enthalpy plus rest mass density of the thermal and cosmic ray gases. The comoving frame gas and cosmic ray momentum equations reduce to

$$\gamma A \frac{dP_g}{dr} + \frac{d}{dr}(A\gamma)W_g = \bar{\nu} T_c''^{10}, \tag{3.17}$$

$$\gamma A \frac{dP_{c\|}}{dr} + \frac{d}{dr}(A\gamma)W_c + \frac{2A\gamma}{r}(P_{c\|} - P_{c\perp}) + \gamma\beta c A \frac{dT_c''^{10}}{dr}$$
$$+ \left[\frac{2\gamma\beta A}{r} + 2\frac{d}{dr}(\gamma\beta A)\right]cT_c''^{10} = -\bar{\nu}T_c''^{10}. \tag{3.18}$$

Assuming the Eddington factor $e = e(r)$ is a given function of r, then these equations can be shown to possess singularities where the flow speed β matches either the cosmic ray sound speed β_{cr}, or the thermal gas sound speed $\beta = \beta_s$, where

$$\beta_{cr} = (\gamma_c^* - 1)^{1/2}, \quad \beta_s = \left(\frac{\partial P_g}{\partial \rho}\right)^{1/2} = \left(\frac{\gamma_g P_g}{W_g}\right)^{1/2}, \tag{3.19}$$

and $\rho = E_g + m_0 n c^2$.

The further investigation of these equations remains to be carried out. Questions of interest include: to what extent does the reaction of the cosmic rays on the fluid modify the inflow?; are shocks necessary in the flow and if so, to what extent are cosmic-rays accelerated at these shocks?

4 Concluding Remarks

In this paper we have discussed various aspects of cosmic ray transport and acceleration in astrophysical fluid flows via diffusive particle transport equations; pitch angle dependent transport equations, and hydrodynamical equations.

The diffusive particle transport equation (2.1) gives a convection- diffusion description of particle transport coupled with various energy change processes. These processes include adiabatic energy changes associated with expansions and compressions of the fluid; non-inertial energy changes associated with acceleration of the fluid (this may also be thought of as gravitational redshift); energy changes of particles scattering across shear flows (cosmic ray viscosity); second order Fermi acceleration; and various energy loss processes.

A discussion of energetic particle transport in an accretion flow onto a Schwarzchild black hole was carried out in section 3. The self consistent hydrodynamics of the accretion flow coupled with the hydrodynamics of the cosmic-rays constitutes a complicated nonlinear problem, since the solution of the "pitch angle" evolution equation (3.3) for the cosmic rays must be solved consistent with the hydrodynamics of the system. The equations exhibit singular behavior when the flow speed β matches either of the cosmic ray sound speed (β_{cr}) or the thermal gas sound speed (β_s). Also evident is the requirement that the flow speed approach the speed of light at the Schwarzschild radius (i.e. $\beta \to 1$ as $r \to 2GM/c^2$).

The adiabatic compression of the cosmic-rays in the inflow in general should lead to an adverse cosmic ray pressure gradient that decelerates the inflow. If the flow involves shocks, further energization of the cosmic rays by the first order Fermi mechanism at these shocks will inevitably take place. The model equations are of course simplified versions of more complex models including non-adiabatic radiative losses and gains, heat transport processes and pair creation processes (see e.g. Babul, Ostriker and Mézáros 1989).

The detailed application of cosmic ray transport equations to the shock acceleration of energetic electrons in radio-jet shocks including the effects of synchrotron losses has been carried out by a number of authors (e.g. Webb, Drury and Bierman 1984; Bregman 1985; Heavens and Meisenheimer 1987; Meisenheimer and Heavens 1986; Fritz 1989a,b; Fritz and Webb 1990; Webb and Fritz 1990; Mitras 1990). These results appear to be of relevance in describing the emission from radio wavelengths to the X-ray regime in AGN's and jets (e.g. Meisenheimer et al. 1989; Biermann and Strittmatter 1987). Other applications include cosmic ray acceleration in supernova shocks (e.g. Dorfi 1990; Drury et al. 1989; Jones and Kang 1990) and particle acceleration in ultrarelativistic shocks (e.g. Kirk and Schneider 1987; Ellison, Jones and Reynolds 1990; Kirk 1990, these proceedings).

Acknowledgements

This work was supported in part by NASA grant NSG 7101 and by NSF grant ATM 8317701. I wish to thank John Kirk for discussions on cosmic ray transport in accretion flows and Sandra Marchello for wordprocessing the manuscript.

References

Achterberg, A. 1981, *Astr. Ap.*, **97**, 259.
Babul, A., Ostriker, J.P. and Mézáros P., 1989, *Ap. J.*, **347**, 49.
Baring, M.G. and Kirk, J.G. 1990, *Astr. Ap.*, in press.
Berezhko, E.G. and Krimsky, G.F. 1981, *Soviet Astr. Letters*, **7**, (5), 352.
Berezhko, E.G. and Krimsky, G.F. 1988, *Soviet Phys. Usp.*, **31**, (1), 27.
Biermann, P. and Strittmatter, P.A. 1987, *Ap. J.*, **322**, 643.
Blandford, R.D. and Eichler, D. 1987, *Phys. Reports*, **154**, 1.
Bregman, J.N. 1985, *Ap. J.*, **288**, 32.
Bridle, A.H. and Perley, R.A. 1984, *Ann. Rev. Astron. Astrophys.*, **22**, 319.
Dolginov, A.Z. and Toptygin, I.N. 1966, *Bull. Acad. Sci USSR, Phys. Ser.*, **30**, 1852.
Dorfi, E.A. 1990, *Ast. Ap.*, **234**, 419.
Drury, L. O'C. 1983, *Rept. Prog. Phys.*, **46**, 973.
Drury, L. O'C, Markiéwicz, W.J., and Völk, H.J., 1989 *Astr. Ap.*, **255**, 179.
Earl, J.A., Jokipii, J.R. and Morfill, G.E. 1988, *Ap. J. (Letters)*, **331**, L91.
Ellison, D.C., Jones, F.C. and Reynolds, S.P. 1990, *Ap. J.*, **360**, 702.
Fritz, K.D. 1989a, *Astr. Ap.*, **214**, 14.
Fritz, K.D. 1989b, *Ap. J.*, **347**, 692.
Fritz, K.D. and Webb, G.M. 1990, *Ap. J.*, **360**, 387.
Gleeson, L.J. and Axford, W.I. 1967, *Ap. J. (Letters)*, **149**, L115.
Hakim, R. 1967, *J. Math. Phys.*, **8**, 1379.
Heavens, A.F. and Meisenheimer, K. 1987, *M.N.R.A.S.*, **225**, 335.
Jones, T.W. and Kang, H. 1990, *Ap. J.*, **363**, 499.
Kirk, J.G. and Schneider P. 1987, *Ap. J.*, **315**, 425.
Kirk, J.G. 1988, *Ap. J.*, **324**, 557.
Kirk, J.G., Schlickeiser, R. and Schneider, P. 1988, *Ap. J.*, **328**, 269.
Krülls, W.M. 1990, *Astr. Ap.*, submitted.
McKenzie, J.F., Ip, W.H. and Axford, W.I. 1979, *Ap. Space Sci.*, **64**, 183.
Meisenheimer K. and Heavens, A.F. 1986, *Nature*, **323**, (2), 419.
Meisenheimer R., Röser H.J., Hiltner P., Yates M.G., Longair M.S., Chini R. and Perley R.A 1989, *Astr. Ap.*, **219**, 63.
Melrose D.B. 1980, *Plasma Astrophysics: Nonthermal Processes in Magnetized Plasmas*, (London: Gordon and Breach), **2**, Chapter 8.
Mitras, A. 1990, *Ap. J.*, **348**, 221.
Ostrowski, M. 1990, *Astr. Ap.*, in press.
Parker, E.N. 1965, *Planet. Space Sci.* **13**, 9.
Riffert, H. 1986, *Ap. J.*, **310**, 729.
Skilling, J.A. 1975, *M.N.R.A.S.*, **172**, 557.
Webb, G.M. 1985, *Ap. J.*, **296**, 319.
Webb, G.M. 1987, *Ap. J.* **319**, 215.

Webb, G.M. 1989, *Ap. J.*, **340**, 1112.

Webb, G.M. 1990, *Proc. 21st Int. Cosmic Ray Conf., Adelaide, Australia*, 4, 126.

Webb, G.M., Drury, L. O'C. and Biermann, P., 1984, *Astr. Ap.*, **137**, 185.

Webb, G.M. and Fritz, K.D. 1990, *Ap. J.*, **362**, 419.

Webb, G.M. and Jokipii, J.R. 1990, *Proc. 21st Int. Cosmic Ray Conf., Adelaide, Australia*, **4**, 122.

Particle Acceleration in Relativistic Shock Waves with Oblique Magnetic Fields in the Presence of Finite Amplitude Field Perturbations

M. Ostrowski

Astronomical Observatory, Jagiellonian University, ul.Orla 171,
30-244 Kraków, Poland

Abstract: Non-monotonic changes of the accelerated particle spectral index and a decreasing particle density jump at the shock are shown to occur in relativistic shock waves with oblique magnetic fields for increasing amplitude of magnetic field perturbations. Very steep particle energy spectra can be produced at such shocks.

1 Introduction

Relativistic shock waves are supposed to occur in a number of situations where relativistic plasma flow velocities are observed. If the flow direction does not coincide with the magnetic field, such a shock is called an "oblique" one. Here, for the cosmic ray particles which move predominantly along the field, the shock velocity U_1 is increased by the factor $1/\cos\alpha_1$, where α_1 is the angle between the shock velocity and the magnetic field (index 1 denotes quantities measured in the upstream plasma rest frame). Kirk and Heavens (1989) described the first-order Fermi CR acceleration process at "sub-luminal" oblique shock waves ($U_1/\cos\alpha_1 < 1$, in units of the light velocity). They considered shocks propagating in the medium with the magnetic field sufficiently uniform to allow for using particle magnetic moment conservation for particle interactions with the shock. As a result, a possibility of very flat particle spectra - i.e. with the spectral index s of the integral distribution function $N(> p)$ close to 0 - has been revealed, due to a very high reflection probability for anisotropically distributed upstream particles. In order to examine the question how this result is affected by the presence of magnetic field perturbations we performed particle simulations involving such perturbations.

2 Particle Acceleration in the Presence of Magnetic Field Perturbations

The Monte Carlo simulations applied in this section are described in more detail in Ostrowski (1990b). Between discrete orbit perturbations particle trajectories were derived for the uniform background magnetic field in the respective local plasma rest frame. At the scattering, the particle momentum vector was scattered in the uniform way within a small cone about the original direction (Kirk and Schneider 1987; Ostrowski 1990a,b). After injecting particles at the shock, we followed their trajectories until they crossed the shock or were advected far downstream, reaching a free boundary there. The escaping particles were registered, with their respective weights, to build up the energy spectrum. At the same time, trajectories of some particles crossing the shock were split (with their weights) in order to keep the total number of particles in the simulations constant. We also recorded particle residence times in spatial bins in order to obtain the particle spatial distribution. The times between used for particle acceleration (as measured in the upstream rest frame) and the corresponding energy gains enabled us to determine the characteristic acceleration time scale t_{acc} for particles of momentum p. In deriving t_{acc} we could use particles of all momenta by setting the acceleration conditions in such a way that $t_{acc}(p) = t_{acc,0} \cdot p/p_0$. We obtained the energy and spatial distributions for a number of situations with different perturbation amplitudes, characterized with

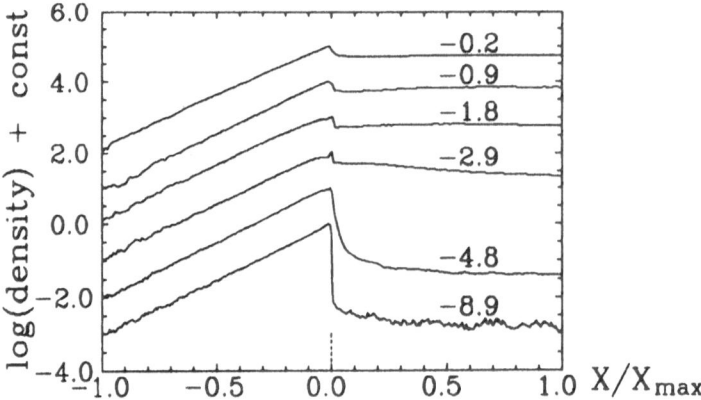

Fig. 1. Particle spatial distributions for a number of $\log D_\perp / D_\parallel$ values, the respective values are given near the curves. Each distribution was given in the shock rest frame and it was normalized to the maximum value at the shock. The distance from the shock where the upstream particle density was 10^{-3} of the maximum was chosen as the scale X_{max}. The shock parameters were $U = 0.5$, $R = 5.11$ and $\alpha_1 = 55°$ ($U_1 / \cos \alpha_1 = 0.87$).

$\eta \equiv D_\perp/D_\parallel$, the ratio of the perpendicular (to the mean magnetic field) diffusion coefficient to the parallel one. The value of η was modified by changing the mean time between scatterings, not the scattering amplitude. In Fig.1 we show the particle spatial distributions for a number of η values for the shock propagating in a cold (e,p) plasma (Heavens and Drury 1988). In Fig.2 the changes of the particle spectral index σ and, in Fig.3, values of $t_{acc,0}$ are presented.

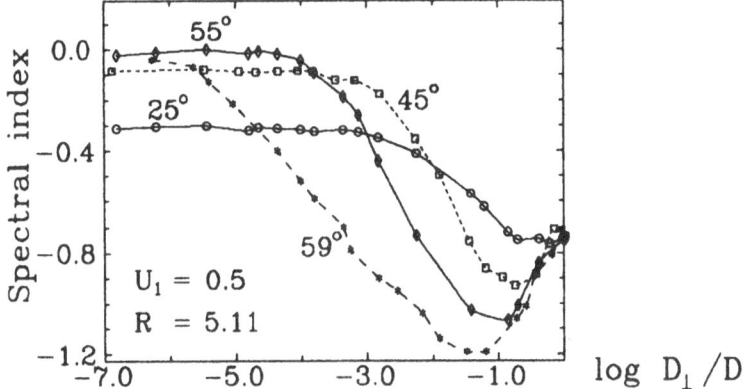

Fig. 2. Particle spectral index σ as a function of $\log D_\perp/D_\parallel$, for a number of magnetic field inclinations α_1. The inclination angles are given near the respective curves.

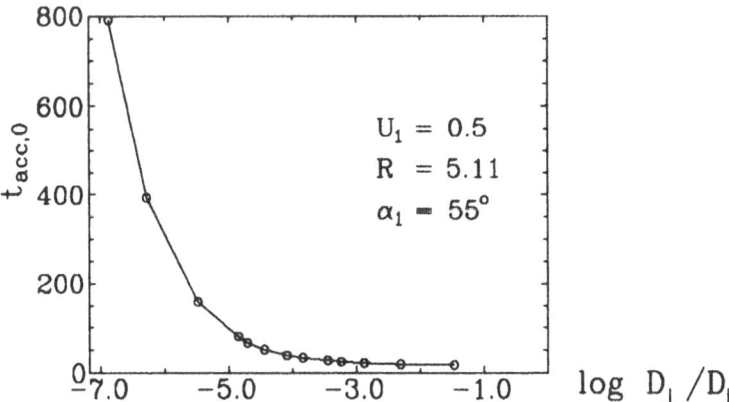

Fig. 3. The characteristic acceleration time $t_{acc,0}$ as a function of $\log D_\perp/D_\parallel$. The values are given in the upstream rest frame in the units of particle gyroradius $r_{g,0}$ over c.

3 Discussion

The results of Kirk and Heavens (1990) are valid for some idealized shock conditions. The existence of perturbed magnetic field (MHD waves) leads to changing the effective magnetic field inclination. As shown in Fig.2, small amplitude perturbations may even increase this inclination, leading to a flatter spectrum, but as their amplitudes grow higher the spectrum becomes steeper, and after reaching a minimum, flattens back to the parallel shock value. These non-monotonic changes are due to the fact that the small amplitude perturbations influence the shock-particle interaction more effectively , increasing the upstream-downstream transmission probability and decreasing the mean energy gain, than the escape probability for downstream particles. The density jump at the shock decreases and, due to the shorter diffusive scale and higher mean energy gain, the characteristic acceleration time also decreases.

The small-amplitude conditions corresponding to the very flat spectra seem to be unstable beyond the test particle limit. Due to the particle-wave interactions, particle streaming generates high-amplitude waves in the shock vicinity. At the same time a particle density jump gives rise to instabilities at the shock. It may have far-reaching consequences for the particle spectra in relativistic shocks, leading in a natural way to steep spectral indices. We expect the analogous situation can hold for quasi-parallel shocks with high amplitude waves, as they exhibit some characteristics of oblique shocks (Ostrowski 1988).

I am grateful to John Kirk and Alan Heavens for their critical remarks and useful comments.

References

Kirk, J.G., Heavens, A.F. (1989): Mon.Not.Roy.astr.Soc. **239**, 995

Kirk, J.G.,Schneider, P. (1987): Astrophys.J. **322**, 256

Heavens, A.F., Drury, L.O'C. (1988): Mon.Not.Roy.astr.Soc. **235**, 997

Ostrowski, M. (1988): Astron.Astrophys. **206**, 169

Ostrowski, M. (1990a): in Proc. 1-st COSPAR Coll. "Physics of Outer Heliosphere", eds. S. Grzedzielski, D.E.Page; Pergamon Press

Ostrowski, M. (1990b): Mon.Not.Roy.astr.Soc. (submitted)

Collisional Acceleration of Electrons in Central Regions of Active Galactic Nuclei

Matthew G. Baring

[1] Max-Planck-Institut für Astrophysik, Karl Schwarzschild Strasse 1, D-8046 Garching bei München, Fed. Rep. of Germany.

Abstract: Coulomb collisions between relativistic protons and cold electrons are investigated to determine the resulting electron excitation spectrum. The cold electrons could be the end products of a pair cascade, such as is used in models of the X-ray and gamma-ray AGN continuum, while the protons are injected into the central region of an AGN by some acceleration process. Large-angle Coulomb scattering is very efficient at producing relativistic electrons, and can be a significant supply of non-thermal electrons for typical AGN conditions. If proton cooling in the collisions is included, very flat electron spectra can be obtained, otherwise the electron injection has an index of at least two. This collisional acceleration may be important for AGN pair cascade models.

1 Introduction

The extensive study of pair cascade models of the X-ray and gamma-ray emission from the central regions of active galactic nuclei (AGN) over the last few years has yielded much understanding about the emission mechanism of AGNs and the interplay of various physical processes that probably arise in compact cosmic objects (e.g. see Svensson, 1987; Lightman and Zdziarski, 1987; and Coppi, 1991, among many others). The literature on this subject has focussed on the spectra obtainable from the models and their comparison with observational data, making important spectral predictions that await confirmation. All of these models assume an arbitrary (but carefully chosen) injection of relativistic electrons into the emission region, ignoring the physics of such an injection. This step is taken to avoid the complexities of particle acceleration processes and also because the understanding of the physical environment of the acceleration region is very uncertain. However, the investigation of possible injection mechanisms is now becoming more salient.

---

[1] Present address: Department of Physics, North Carolina State University, Raleigh, NC 27695, U.S.A.

The direct acceleration of electrons in the emission region, either through shock acceleration (which may be quite inefficient) or electrodynamic acceleration in poorly-understood turbulent fields above an accretion disc, is not currently popular. More favoured is the idea where protons provide the electron injection indirectly via hadronic interactions (e.g. Kazanas and Ellison, 1986; Sikora et al., 1987). Protons can be efficiently accelerated by shocks to large Lorentz factors and also can have mildly relativistic velocities in spherical accretion flows. Another mechanism for the indirect injection of electrons into the compact emission region of AGNs is the collisional acceleration of electrons by protons. This process, which has not been considered in detail for electron injection in AGNs before, is the subject of this paper. Here, the conditions of the central regions of AGNs are found to be quite favourable for the production of a high luminosity of relativistic electrons. Energetic protons cool mostly by extremely small-angle Coulomb collisions with cold electrons that arise naturally in pair cascade models, while a small proportion of collisions involve "large-angle" scatterings, which provide relativistic boosts to the electron. It is these large-angle Coulomb collisions that provide a principal source of electron injection.

2 Scattering Kinematics

Relativistic protons colliding with cold electrons cool mostly by small-angle collisions (yielding the Coulomb logarithm in the cooling rate). However, a small proportion of scatterings are of large angle and produce relativistic electrons, contributing of the order of a few percent to the proton cooling rate (Baring, 1991). These large angle collisions occur with a cross-section of the order of σ_T, the Thomson cross-section, and can be very efficient in producing relativistic electrons in AGN (see Eq. (8)). In this paper it is convenient to use dimensionless units, scaling all particle energies by $m_e c^2$. Further, it is assumed that $c = 1$. The electron and proton energies before collision are $\varepsilon_e \approx 1$ and $\varepsilon_p = \gamma_p m_p$, respectively, for a proton Lorentz factor γ_p, where $m_p/m_e \approx 1836$.

The consideration of the collisional kinematics of $ep \rightarrow ep$ elucidates the understanding of this interaction as a means of electron acceleration. Of particular interest is the maximum energy that the electron can attain as a result of a single scattering. Since the laboratory frame is inconvenient for the examination of collisional spectra for the production of electrons, it is expedient to choose variables in a special frame of reference, in which the scattering cross-section is known. Here, the centre-of-momentum (CM) frame is preferred, and the conventions and notation of Baring (1987) will be adopted. Assuming an electron to be almost at rest in the laboratory frame, if it collides with a proton of much larger momentum, the Lorentz transformation from the laboratory frame to the CM frame is characterized by the boost $\beta_c = p_p/(e_p + m_e)$. Let ε_e^* and ε_p^* be the electron and proton energies in the CM frame, respectively, and p^* be the (common) magnitude of their momenta in this frame. Then if they move after collision at an (CM) angle Θ^* with respect to the Lorentz boost vector, and defining $\mu^* = \cos \Theta^*$, the final electron and proton energies in the *laboratory* frame are (Baring, 1987):

$$\varepsilon'_e = \gamma_c(\varepsilon^*_e - \mu^* \beta_c p^*) \quad , \qquad \varepsilon'_p = \gamma_c(\varepsilon^*_p + \mu^* \beta_c p^*) \quad , \tag{1}$$

where $\gamma_c = 1/\sqrt{1 - \beta_c^2}$ is the gamma-factor for the Lorentz boost between the two frames. The case of $\mu^* = 1$ reproduces the initial conditions ($\varepsilon'_e = \varepsilon_e$, $\varepsilon'_p = \varepsilon_p$). It can easily be deduced from equation (1) and the initial conditions, that a scattering specified by μ^* in the CM frame yields a final electron energy ε'_e given by (Baring, 1991):

$$\frac{\varepsilon'_e}{m_e} = \frac{(\varepsilon_p + m_e)^2 - \mu^* p_p^2}{2\varepsilon_p m_e + m_e^2 + m_p^2} \quad . \tag{2}$$

The $\mu^* = 1$ case yields $\varepsilon'_e = m_e$, but observe that relativistic boosts of the electron's energy are achievable for small scattering angles ($\mu^* \approx 1$) when the incident proton is relativistic. More specifically, the maximum value of the electron's energy after a collision, which occurs for backscattering in the CM frame ($\mu^* = -1$), has the following limiting values that are obtainable from equation (2):

$$\left. \frac{\varepsilon'_e}{m_e} \right|_{\max} \approx \begin{cases} 2\gamma_p^2 - 1, & 1 \leq \gamma_p \ll m_p/m_e ; \\ \gamma_p m_p/m_e, & \gamma_p \gg m_p/m_e . \end{cases} \tag{3}$$

For low proton energies, namely $\gamma_p \ll m_p/m_e$, the scattered electrons take only a small fraction of the proton's initial energy while for $\gamma_p \gg m_p/m_e$ the scattering is discrete. Clearly, mildly-relativistic protons can only accelerate electrons to mildly relativistic energies. Therefore, a thermal supply of protons such as would be provided in disc or spherical accretion models of AGNs could not yield the relativistic electron injection that is needed for AGN X-ray and γ-ray spectral modelling via this collisional mechanism. Relativistic protons can, however, produce desirable electron energies. If $\gamma_p \sim 10$, the electrons are accelerated up to $\gamma_e \sim 100$, the value that proves ideal for the model of Done et al. (1990). They demonstrate that such Lorentz factors provide the best conditions to produce AGN spectra without the optical depth of cold electrons being too large. The shock acceleration of protons to extremely large energies ($\gamma_p \sim 10^8$: see Sikora et al., 1987) would yield electron Lorentz factors as large as 10^{11} after collision.

3 Electron Excitation Spectra

The collisional excitation spectra of electrons are naturally of great relevance to any pair cascade model of AGN spectra. The spectrum of electrons accelerated by Coulomb collisions with relativistic protons is calculated here using the well-known Rutherford cross-section, which considers only scatterings between spinless point electrons and protons and neglects the proton recoil. This is the simplest form for scattering: a more exact treatment is given in Baring (1991) using the more general Dirac and Rosenbluth cross-sections (e.g. see Perkins, 1982). The Dirac cross-section is an extension of the Rutherford formula to include the particles' spins and the collisional momentum transfer (i.e. proton recoil). The assumption that the colliding particles are point charges is relaxed in the Rosenbluth cross-section, which includes the form factors describing the electromagnetic "charge" distributions of the electron and the proton, and is the most general Coulomb cross-section.

Since the CM frame is chosen here to describe the angle of scattering, the well-known Rutherford result (e.g. see Jauch and Rohrlich, 1980) must be Lorentz transformed from the proton rest-frame. This is easily performed by standard techniques (see Baring, 1991; and also Stepney, 1983) to yield the Rutherford scattering differential cross-section in the CM frame, valid for small scattering angles:

$$\frac{d\sigma^*}{d\mu^*} \approx \frac{2\pi r_o^2}{(1-\mu^*)^2} \frac{m_p^2 \varepsilon_p^2}{p_p^4} \left[1 + \frac{2m_e \varepsilon_p}{m_p^2}\right] . \tag{4}$$

This result specializes to equation (4.5) and (4.6) of Stepney (1983). The spectral rate of production of electrons, differential in energy, is then simply this cross-section multiplied by the relative velocity in collisions and the factor (obtained from Eq. (2)) $d\mu^*/d\varepsilon = (2\varepsilon_p m_e + m_e^2 + m_p^2)/(m_e p_p^2)$:

$$\frac{dn_e(\varepsilon)}{dt} = n_e n_p \beta_{rel} \frac{d\mu^*}{d\varepsilon} \frac{d\sigma^*}{d\mu^*} . \tag{5}$$

Here the relative velocity is $\beta_{rel} = p_p/\varepsilon_p$ and μ^* is evaluated using equation (2) with $\varepsilon_e' = \varepsilon$. The simple dependence of the differential cross-section on μ^* and the linear parameterization of ε by μ^* in equation (2) yield a very simple energy dependence in the electron's acceleration spectrum for monoenergetic protons if $\varepsilon \gg m_e$:

$$\frac{dn_e(\varepsilon)}{dt} \approx \frac{3n_e n_p \sigma_T}{4m_e \varepsilon^2} \frac{\varepsilon_p}{p_p} , \tag{6}$$

where $\sigma_T = 8\pi r_o^2/3$ is the Thomson cross-section. This simple form does not arise when the more complicated Rosenbluth formula is used (Baring, 1991), since the cross-section progressively decreases below the Rutherford result as $1 - \mu^*$ becomes significantly large.

The spectral slope of the electron excitation spectrum changes when a broad distribution of protons is present. Protons are very probably injected with a power-law distribution, particularly if they are energized via the shock acceleration mechanism. With this in mind, the distribution $n_p(\gamma_p) = n_p (\Gamma - 1)\gamma_p^\Gamma$ is considered, for which the produced electron spectrum is (using equations (3) and (6)):

$$\frac{dn_e(\varepsilon)}{dt} \approx \frac{3n_e n_p \sigma_T}{4m_e} \begin{cases} \left(\frac{\varepsilon}{2m_e}\right)^{-(3+\Gamma)/2} , & \varepsilon \ll m_p^2/m_e ; \\ \left(\frac{\varepsilon}{m_p}\right)^{-(1+\Gamma)} , & \varepsilon \gg m_p^2/m_e ; \end{cases} \tag{7}$$

If $\Gamma \geq 1$, the flattest electron spectrum that can be obtained has an index of two. This would prove to be a major constraint on AGN spectral models if this collisional excitation process forms the dominant electron injection mechanism. Note also that for $\Gamma > 1$, the spectra break above $\varepsilon \sim m_p^2/m_e$. However, if the protons are injected into the emission region with a power-law of index $\Gamma \geq 1$, and cool predominantly through Coulomb collisions with cold electrons, the steady-state cooling distribution of the protons is a power-law of index $\Gamma - 1$ (Baring, 1991). This is because the proton cooling rate $d\gamma_p/dt$ is roughly independent of γ_p. Then the electron spectra can have an index as low as unity. Other cooling processes for protons are considered by Begelman, Rudak and Sikora (1990), of which Coulomb collisions with cold protons is the dominant source of cooling for protons with relatively low gamma-factors (certainly for $\gamma_p < m_p/m_e$). Only for high densities of cold protons can the proton-proton cooling compete with ep collisions (Baring, 1991). Proton-photon collisions are the dominant form of proton cooling for very high energies only (Sikora, et al. 1987) so that ep cooling of the protons may indeed yield flat electron injection distributions. Note also that although Coulomb collisions with cold thermal electrons can cool relativistic electrons in AGN pair cascades (Done, et al., 1990; Coppi, 1991), this does not provide a significant reinjection of electrons. This is because the non-thermal electron density in a pair cascade is quite low due to their rapid Compton cooling by UV photons.

It is also appropriate to outline the conditions for ep Coulomb scattering to be important in the central region of an AGN. The injection luminosity of electrons L_e can be simply calculated, for example in the case of monoenergetic protons, integrating equation (6) times ε over energies ε. This is most conveniently expressed in terms of the dimensionless luminosity or compactness parameter $l_e = L_e \sigma_T/(Rm_e)$ of previous pair cascade studies (e.g. see Lightman and Zdziarski, 1987). For an optical depth $\tau_T = n_e \sigma_T R$ of cold electrons with $\tau_p = n_p \sigma_T R$ describing the proton density, in an AGN region of size R (i.e. volume $4\pi R^3/3$), equation (6) yields a dimensionless luminosity of injection of

$$l_e \sim \pi \tau_T \tau_p \log \varepsilon_{max} \quad . \tag{8}$$

Here ε_{max} is the maximum electron energy (see Eq. (3)). This electron injection compactness parameter is normally of the order of unity or greater in pair cascade models. Pair cascades usually yield τ_T of the order of unity or greater (Lightman and Zdziarski, 1987), so that it can be concluded that when $\tau_p \sim 1$, $l_e \gtrsim 1$ and ep collisional excitation of electrons is an important source of electron injection in AGN models. Achieving $\tau_p \sim 1$ is not that difficult, particularly since protons might be present in the emission region with large density because of charge neutrality. Having an injection (or cascade reacceleration) of *pairs* reduces τ_p. Plausible particle acceleration mechanisms can easily inject protons into the emission region with the required densities. Therefore Coulomb collisions can obviously be prominent in AGN models. Finally note that ep bremsstrahlung collisions can also be a significant source of electron reinjection in AGNs when $\gamma_p \gtrsim 100$ (Baring, 1991).

4 Summary

Large angle electron-proton Coulomb scattering is found to provide a significant source of relativistic electron injection in the central regions of AGNs if the density of relativistic protons is such that $\tau_p = n_p \sigma_T R \gtrsim 1$. This competes with other modes of injection. Mildly-relativistic protons can produce only mildly-relativistic electrons. For monoenergetic protons, the electron injection has an energy dependence of ε^{-2}, while for power-law protons, flatter injections can be obtained if the protons cool predominantly through these collisions. This electron reacceleration mechanism may have to be incorporated in AGN pair cascade models that include particle acceleration self-consistently.

References

Baring, M. G.: 1987, *M.N.R.A.S.* **228**, 681
Baring, M. G.: 1991, *M.N.R.A.S.*, to be submitted
Begelman, M. C., Rudak, B., and Sikora, M.: 1990, *Ap. J.* **362**, 38
Coppi, P. S.: 1991, *M.N.R.A.S.*, in press
Done, C., Ghisellini, G., and Fabian, A. C.: 1990, *M.N.R.A.S.* **245**, 1
Jauch, M. M., and Rohrlich, F.: 1980, *The Theory of Photons and Electrons* (2nd Edition; Springer, Berlin)
Kazanas, D., and Ellison, D. C.: 1986, *Ap. J.* **304**, 178
Lightman, A. P., and Zdziarski, A. A.: 1987, *Ap. J.* **319**, 643
Perkins. D. H.: 1982, *Introduction to High Energy Physics* (2nd Edition; Addison-Wesley, Reading, Mass.)
Sikora, M., Kirk, J. G., Begelman, M. C., and Schneider, P.: 1987, *Ap. J. (Lett.)* **320**, L81
Stepney, S.: 1983, *M.N.R.A.S.* **202**, 467
Svensson, R.: 1987, *M.N.R.A.S.* **227**, 403

Influence of Second-Order Fermi Acceleration and Relativistic Shock Waves on Nonthermal Continuum Emission in Hot Spots

Wolfram M. Krülls

Institut für Theoretische Kernphysik der Universität Bonn
Nußallee 14–16, D-5300 Bonn 1, Fed. Rep. of Germany
and
Max-Planck-Institut für Radioastronomie, Auf dem Hügel 69
D-5300 Bonn 1, Fed. Rep. of Germany

Abstract: We have investigated modified descriptions of synchrotron spectra in hot spots beyond the standard model of diffusive shock acceleration. We have considered second order Fermi acceleration in combination with modified shock fronts. In addition the role of relativistic shock waves in hot spots is re-examined. With these modifications the following results are obtained. In general it is always possible to get the same shape of synchrotron spectra as in the standard model of diffusive acceleration by pure second-order Fermi acceleration. Deviations of the shape of the distribution functions occur only near the injection momentum or for large Alfvén velocities. Hence it is possible to have combined spectra of first and second-order Fermi acceleration in hot spots. Due to the possibility of a power-law index $\alpha > -0.5$, flat spectra (3C273) can be explained. In contrast, the relativistic shock waves are not able to generate cutoffs which are steep enough to fit the observed spectra. The reason is the fact that we usually only see the part of the anisotropic spectrum in optically thin sources directed towards us and therefore at small pitch angles. But at small pitch angles cutoffs are less steep than in the diffusive acceleration model. From this we can conclude that the shock wave velocities in hot spots seem to be smaller than $0.5c$.

1. The Diffusion Approximation

The photometry of hot spots can be an important test for theories of acceleration processes if we know about the nonthermal origin of the received optically thin radiation. Recent photometry of some hot spots and jets (Meisenheimer et al. 1989) shows polarization, similar at different wavelengths confirming the nonthermal nature of the observed emission.

131

We assume the following physical situation: There is an anisotropic thermal background medium with a magnetic field B_0 and Alfvén turbulence with waves parallel to this field. Let the Alfvén velocity be small against the velocity of the energetic particle component which we are interested in. We do not take into account the back-reaction of these energetic particles with the turbulent Alfvén wave field. Furthermore we assume the identity of the intensities of left- and right-polarized waves streaming forward and backward in the background medium and the change in streaming velocity $\beta(z)$ is parallel to the magnetic field (parallel shock wave).

Let the quickest process be scattering in pitch-angle so that we can use the diffusion approximation. Then with all these conditions the transport equation in the stationary case reduces for single-charged particles to

$$c\beta\frac{\partial \bar{f}}{\partial z} = \frac{\partial}{\partial z}\left(\kappa\frac{\partial \bar{f}}{\partial z}\right) + \frac{c}{3}p\frac{d\beta}{dz}\frac{\partial \bar{f}}{\partial p} + \frac{1}{p^2}\frac{\partial}{\partial p}\left(a_2p^2\frac{\partial \bar{f}}{\partial p}\right) + \frac{2}{3}\frac{k_{syn}}{p^2}\frac{\partial}{\partial p}\left(p^4\bar{f}\right) + Q \,,$$

with

$$k_{syn} = \frac{1}{m_0ct_{syn}} = \frac{e^4B_0{}^2}{6\pi\varepsilon_0m_0^4c^4} \,; \qquad Q = \frac{n_0}{4\pi p_0^2}\delta(z-z_0)\delta(p-p_0) \,,$$

where p is the absolute value of the energetic particle's momentum, c the velocity of light, κ and a_2 the spatial and momentum diffusion coefficients, and $\bar{f}$ is the pitch-angle averaged one-particle distribution function. We shall use a smooth velocity profile given by a hyperbolic tangent function due to the unimportance of the exact profile across the shock wave.

To solve this differential equation we have used a Galërkin spectral method which is semi-analytical. We make an ansatz with Čebyšev polynomials of the second order. The ansatz function fulfils the boundary conditions in z. Using such a trial function in z and the same one as a test function we get with the recurrence relations of the polynomials a system of ordinary differential equations. The larger this system of differential equations, i.e. N, the better the approximation of the true solution becomes. The system of differential equations will then be solved by a relaxation method.

In Fig. 1 we can see the resulting distribution function in dependence of the parameter $\alpha_1 = \kappa a_2/p^2c^2$ which is equivalent to the characteristic length scale of the considered object divided by the product of the characteristic diffusion times in space and momentum. It is independent of momentum. If α_1 is larger than one second-order Fermi acceleration is the dominant process over first-order acceleration, and vice versa. We see that if the influence of the second-order process becomes dominant a pile-up arises near the cutoff. The general features of the distribution function, however, are the same as in the standard model of diffusive acceleration (Webb et al. 1984). There is first of all a power law, and secondly a cutoff with a steepness independent of α_1. The difference is in the power-law index, which can become larger than -4 (the maximum is -3). The influence of the shock acceleration can be seen in Fig. 2 for different upstream streaming velocities. Because the change in power-law index and cutoff momentum is almost the same for increasing the velocity from zero to a quarter and then to half the speed of light the efficiency of acceleration is growing (look in (Kirk and Schneider 1987) for the validity range of the diffusive model). In Fig. 3 we can see the comparison of a model spectrum of combined first and second-order Fermi acceleration with an observed hot spot spectrum. We see that it is possible to fit the observations by such a combined spectrum. This means that without having strong shock waves it is possible to get model spectra which show a power-law index close to -0.5.

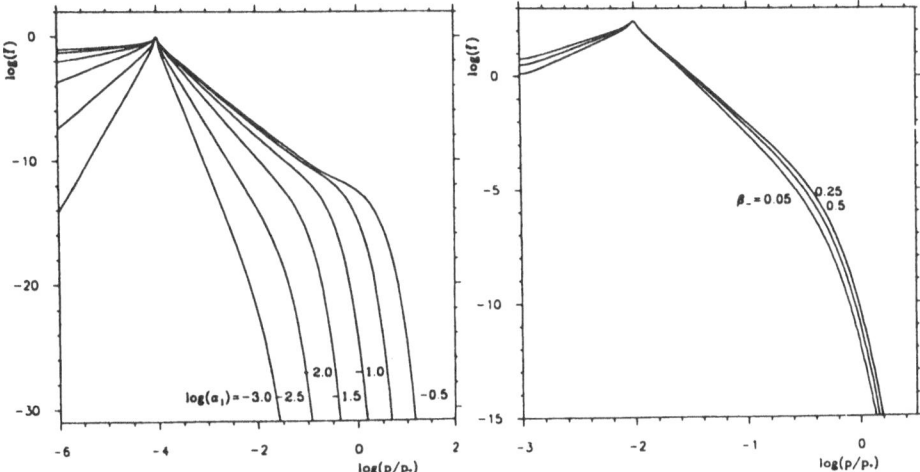

Fig. 1. Normalized distribution functions for different values of the parameter α_1 and without shock wave. For decreasing α_1 the cutoff momentum decreases.

Fig. 2. Distribution function for three different upstream velocities with a constant compression ratio of 4 and $\alpha_1 = 0.05$.

2. Relativistic Shock Waves

Furthermore we have considered the case of relativistic shock waves. For that we solve the general transport equation with synchrotron losses on either side of the shock wave. Because of difficulties solving this problem by numerical methods we use a Monte-Carlo simulation of the transport process. Special methods to increase the variance are used, viz. splitting and efficiency weighting following the way (Kirk and Schneider 1987) handle this problem.

The computations show two important deviations from the distribution function given in the diffusion approximation. Firstly at velocities higher than $0.5c$ the cutoff of the distribution function is going to higher momenta proportional to a power of the Lorentz factor of the difference between upstream and downstream velocities. Secondly the cutoff becomes less steep with increasing streaming velocities. The pitch-angle dependence of the distribution function yields weak changes

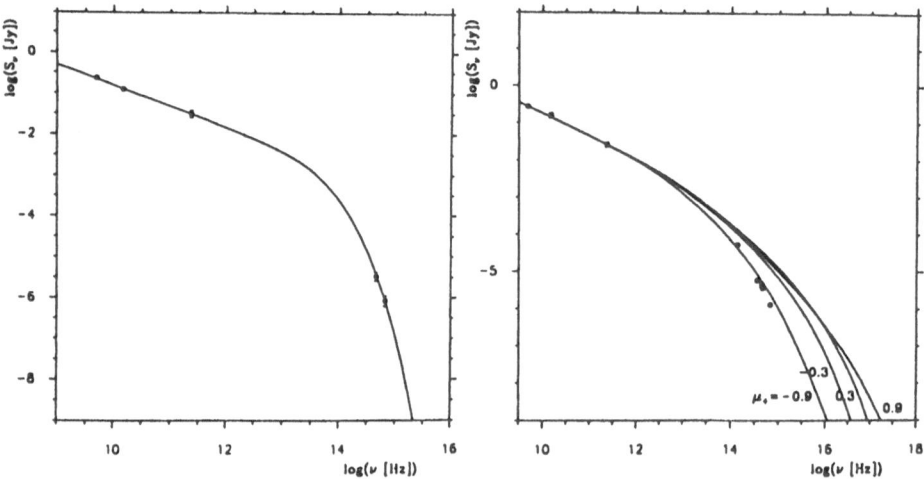

Fig. 3. Spectrum of the western hot spot in 3C20 fitted by a model spectrum with $\alpha_1 = 0.05$, $\beta_- = 0.1$, and $r \approx 3.8$.

Fig. 4. Spectrum of the southern hot spot in 3C33 and spatially integrated model spectra for different pitch angles.

below the pitch angle $\mu_+ < \beta_+$ in the downstream frame. After a maximum a steep drop of the particle density emerges near the forward direction. Additionally the cutoff at high momenta flattens with growing pitch angle. The larger the upstream velocity, the larger the range of pitch angles showing this change in cutoff steepness. Now if the jets in AGN are directed towards us in optically thin sources we can see only the radiation of particles with positive pitch angles. And if the shock waves were relativistic, we would see smoother cutoffs than for the nonrelativistic velocities described in the diffusion approximation. In Fig. 4 we show a comparison of model spectra with different pitch angles for an upstream velocity of $\beta_- = 0.99$ ($r = 3$). It can be seen that it is not possible to fit the observed spectra due to too smooth cutoffs. From these results we can conclude that shock waves in the observed hot spots must be shock waves with upstream velocities smaller than $0.5c$ so that the diffusion approximation is valid as argued in (Meisenheimer et al. 1989).

References

Kirk, J.G., Schneider, P. 1987, *Ap. J.*, **322**, 256
Meisenheimer, K., Röser, H.-J., Hiltner, P.R., Yates, M.G., Longair, M.S., Chini, R., Perley, R.A. 1989, *Astron. Astrophys.*, **219**, 63
Webb, G.M., Drury, L.O'C., Biermann, P. 1984, *Astron. Astrophys.*, **137**, 185

Collimated Relativistic Winds Driven by Electromagnetic Forces

T. Chiueh [1], Z.-Y. Li [2], M. C. Begelman [2]

[1]Institute of Physics and Astronomy, National Central University,
Chung-Li, Taiwan
[2]Joint Institute of Laboratory Astrophysics, University of Colorado
Boulder, CO 80309

Abstract: We report our recent progress on two aspects of electromagnetically driven relativistic winds. First, we show that these winds, as the nonrelativistic counterpart, must always collimate to either current-free paraboloidal jets, or current-carrying cylindrical jets asymptotically. Second, we demonstrate the type of cylindical jets by seeking a self-similar solution, which is a generalization of the self-similar nonrelativistic winds of Blandford and Payne, and the asymptotic structure of which can be shown to be stable against small-amplitude perturbations.

1 Introduction

Superluminal motion observed in the milliarcsecond-resolution VLBI maps of AGN strongly suggests that matters be ejected from the central engine at a relativistic speed (Zensus and Pearson (eds.) 1987). Furthermore, pulsar winds are currently believed to possess a relativistic speed in providing the neutron star spindown energy to the supernova remnants (Rees and Gunn 1974). For stellar winds, non-relativistic bipolar outflows are most commonly observed among young stars and protostars (Lada 1985). Owing to their omnipresence, a majority of these jets might even be driven by a common mechanism.

Here, we explore a possible mechanism suggested by Blandford (1976) and Lovelace (1976) that the flows are driven by rotating magnetic fields. In particular, we will confine ourselves to relativistic flows, which have been investigated before formally by Lovelace et al (1986) and numerically by Camenzind (1989). The issues concern us are why an outflow should collimate to a jet and how a jet gets accelerated to such a high speed. The second issue will be confronted by exploring a particular (self-similar) type of solution.

2 Collimation of Relativistic Winds

It is well-known in the stellar wind literature that relevant MHD equations can be grouped together to yield conservation of angular momentum and of energy along each poloidal field line (Weber and Davis 1967; Goldreich and Julian 1970). They are obtained by considering balance of the force component along the poloidal field line. The two conservation laws can be further combined to yield a so called wind equation, which permits analyses of the criticality conditions of a wind. Although one needs in addition to consider force balance across the poloidal flux surfaces to solve the entire wind problem, the fact that the rotating-field-driven winds must collimate can nevertheless be shown to follow merely from a consideration of the wind equation (Chiueh, Li and Begelman 1990a). (The collimation is defined to be

$$\left(\frac{R}{Z}\right)\bigg|_{Z\to\infty} \to 0,$$

where R and Z are the cylindrical radius and the vertical distance of the poloidal field line trajectory.)

The proof proceeds as follows. The relativistic wind equation reads

$$\left[\frac{\mu}{c^2} - \frac{\frac{l\Omega}{c^2}\left(1 - \frac{\mu\Omega R^2}{lc^2}\right)}{1 - \frac{k^2}{4\pi\rho} - \frac{R^2\Omega^2}{c^2}}\right]^2 = 1 + \left(\frac{kB_p}{4\pi\rho}\right)^2 + \left(\frac{l}{Rc}\right)^2\left[1 - \frac{1 - \frac{\mu\Omega R^2}{lc^2}}{1 - \frac{k^2}{4\pi\rho} - \frac{R^2\Omega^2}{c^2}}\right]^2, \quad (1)$$

where V_p and B_p are the poloidal components of the fluid 4-velocity and the magnetic field, and Ω, μ, l, ρ and k are the angular velocity, specific energy, specific angular momentum, mass density and the ratio of the mass flux to the magnetic flux, respectively. Now, if the poloidal field line diverges away from the rotational axis in such a way that $(Z/R)|_{R\to\infty} = 0$, we will have $B_p R^2 \to \infty$ simply from some geometrical considerations. In this situation, one can easily show that the wind equation can not be satisfied for any value of $V_p|_{R\to\infty}$. Therefore, field lines cannot diverge from the rotational axis, i.e., $(Z/R)|_{R\to\infty} \neq 0$. A similar proof has been given by Heyvaerts and Norman (1989) for nonrelativistic flows recently.

The remaining choices for the flux surface configuration are that it is conical, paraboloidal or cylindrical. We can show that the conical field can neither fill up the polar region, nor match to other types of field at the polar region, and hence it cannot exist (Chiueh, Li and Begelman 1990a). We can further show that the asymptotically cylindrical field carries a net electric current; moreover it must collimate to a cylindrical configuration at a finite distance from the axis, and the kinetic energy flux and Poyning flux are of the same order of magnitude. For the asymptotically paraboloidal field, we can by contrast show that it must carry no net current and the kinetic energy flux must much exceed the Poynting flux. The last point results from the fact that since the asymptotic current diminishes the flux surfaces must diverge rapidly from one to another accordingly, and as a consequence the Poynting flux becomes vanishingly small as compared with the kinetic energy flux. This point is especially important for pular winds, such as the one inside the Crab Nebula. (See Li, Begelman and Chiueh in this volume for a detailed discussion.)

3 Self-Similar Cylindrical Jets

This type of self-similarity arises naturally from dimensional analysis (Chiueh, Li and Begelman 1990b). Velocity $\mathbf{V}$ scales with the speed of light c, the vertical distance Z and cylindrical radius R scale similarly, and mass density ρ scales the same as k^2. The magnetic field $\mathbf{B}$ therefore scales similarly as $\rho\mathbf{V}$. It is important to point out the restriction of this kind of scaling: the angular velocity Ω of the magnetic field, which is presumably anchored to a disk, must scale as c/R and hence deviates from the usual Keplerian rotation law.

The geometry of the self-similarity is best perceived by considering any two poloidal flux surfaces, one enclosing the other. Any straight cone emergent from the origin must intersect each surface. The self-similarity demands the strip of flux surface bounded by the intersection must resemble the other in shape but not in size. The mathematical manifestation of this scaling reads:

$$R = x\left(\frac{\psi}{\psi_0}\right)^\beta , \; Z = y(x)\left(\frac{\psi}{\psi_0}\right)^\beta , \; \mathbf{B} = (b_p(x), b_\phi(x))\left(\frac{\psi}{\psi_0}\right)^{1-2\beta}$$

$$\Omega = \Omega_0\left(\frac{\psi}{\psi_0}\right)^{-\beta} , \; k = k_0\left(\frac{\psi}{\psi_0}\right)^{1-2\beta} , \; \rho = a(x)\left(\frac{\psi}{\psi_0}\right)^{2-4\beta} ,$$

where X and ψ are the two new coordinates, and ψ is called the (poloidal) flux function, related to the poloidal magnetic field by

$$\mathbf{B}_p = \frac{1}{R}\hat{\phi} \times \nabla\psi.$$

(The quantity ψ_0 is some reference flux function, and Ω_0 and k_0 are the corresponding Ω and k at that reference surface.) From a straightforward consideration of poloidal field line geometry, one finds that

$$\mathbf{b}_p = \frac{1}{\beta x(x\dot{y} - y)}\hat{R} + \frac{\dot{y}}{\beta x(x\dot{y} - y)}\hat{Z}. \tag{2}$$

As mentioned before, we must solve the force-balance equation across the poloidal flux surfaces, in addition to that along the flux surface. The latter yields a scaled equation of eq.(1), and the former gives rise to a second-order nonlinear differential equation with a very complicated expression. The equation essentially has a form:

$$\ddot{y} = \frac{N(x, y, \dot{y}, \tau)}{D(x, y, \dot{y}, \tau)}, \tag{3}$$

and

$$D \equiv \left(1 - (1 + \tau)x^2\right) \left[b_\phi^2 + b_p^2 \left(1 - (1 + \tau)x^2\right) + \left(\frac{x + y\dot{y}}{x\dot{y} - y}\right)^2 \left(b_\phi^2 + b_p^2(1 - x^2)\right) \right],$$

where $\tau(x) \equiv k_0^2 c^2 / 4\pi x^2 \Omega^2 a(x)$. This differential equation becomes singular when $D = 0$; the singularity associated with the first factor's being zero is the usual Alfven point and that associated with the second factor is the modified fast point. (The modified fast point occurs when V_θ equals the fast magnetosonic speed, where θ is the polar angle. The explanation for why it should arise has been given by Blandford and Payne (1982) and will not be repeated here.) When passing through the Alfven point, we demand that $N = 0$ as well and use the L'Hopital's rule to calculate $\ddot{y}$ in that neighborhood. The ordinary fast point has no singularity and hence the solution has no trouble in passing through. The modified fast point is a real sigular point, and we numerically find that the valid, nonsingular solutions get to this point only at infinity (Chiueh, Li and Begelman 1990b), much the same as the situation encountering in the radial cold winds (Michel, 1969; Goldreich and Julian, 1970).

Moreover, we find that valid solutions exist not for every value of β but when $0 < \beta < 1$. When $\beta \to 0$, the flux surface asymptotes to a cylinder with radius much greater than the light-cylinder radius, and the ratio of the kinetic energy flux to Poynting flux is maximum, approaching unity. Interestingly, in this limit, the energy output ($\propto \int T^{03} R dR$, where T^{03} is the energy flux in the vertical direction) is the same at every decade of radius.

We also discover that the specific energy μ is a sensitive function of the magnetization parameter, $\sigma \equiv \Omega_0^2 / 4\pi c^3 k_0$, and is numerically found to be $\mu \sim 2\sigma$ for $\sigma \gg 1$. An empirical formula for the terminal Lorentz factor has been numerically determined:

$$\gamma \sim \mu - \frac{\sigma}{\beta} \sim \sigma(2 - \frac{1}{\beta}), \tag{4}$$

when $\sigma \gg 1$. Plotted in Figure 1 is a comparison of two $\gamma(x)$'s with different β's but the same σ. Indeed, the case of larger β requires a greater radial distance for the flow to collimate and it has a larger terminal specific kinetic energy. Remarkably, Figure 1 also shows that major acceleration of the jets occurs in the post-fast-point region, in great contrast to the situations of radial winds.

Finally, we can prove that the interior of the asymptotic structure of the cylindrical jet is stable against small-amplitude perturbations. The proof takes advantage of the fact that the terminal velocity at every flux surface is the same, and hence in reference frame of the jet, the cylindrical equilibrium is a static equilibrium. The usual MHD stability criterion given by the energy principle can be applied to this system (Newcomb, 1963). We find that the equilibrium is stable despite the fact that the magnetic field is dominantly toroidal, an intuitively highly unstable configuration. The reason why it should be stable can be traced to the presence of a singular magnetic field at the axis, according to this self-similar scaling. The strong field provides rigidity to the equilibrium against displacement at the axis, and therefore prevents the most unstable kink mode from occurring.

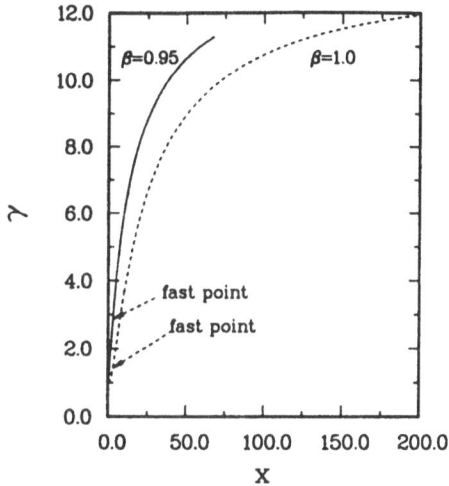

Fig. 1.

We thank supports from grants NSC79-0208-M008-32 of Taiwan and AST88-16140 of NSF, NASA Astrophysics Theory Center grant NAGW-766 and the Alfred P. Sloan Foundation.

References

Blandford, R.D., 1976 *M.N.R.A.S.*, **176**, 465.

Blandford, R.D. and Payne, D.G. 1982, *M.N.R.A.S.*, **199**, 88.

Camenzind, M. 1989, in *Accretion Disks and Magnetic Fields in Astrophysics*, ed. G. Belvedere (Dordrecht: Kluwer), p. 129.

Chiueh, T., Li, Z-Y and Begelman, M.C. 1990a, submitted to *Ap. J. Lett.*

Chiueh, T., Li, Z-Y and Begelman, M.C. 1990b, to be submitted to *Ap. J.*

Goldreich, P. and Julian, W.H. 1970, *Ap. J.*, **160**, 971.

Heyvaerts, J. and Norman, C.A. 1989, *Ap. J.*, **347**, 1055.

Lada, C.J. 1985, *Ann. Rev. Astron. Astrophys.*, **23**, 267.

Lovelace, R.V.E. 1976, *Nature*, **262**, 649.

Lovelace, R.V.E., Mehanian, C., Mobarry, C.M. and Sulkanen, M.E. 1986, *Ap. J. Suppl.*, **62**, 1.

Michel, F.C. 1969, *Ap. J.*, **158**, 727.

Newcomb, W.J. 1960, *Ann. Phys*, **10**, 232.

Rees, M.J. and Gunn, J.E. 1974, *M.N.R.A.S*, **167**, 1.

Weber, E.J. and Davis, L. 1967, *Ap. J.*, **148**, 217.

Zensus, J.A. and Pearson, T.J. (eds.) 1987, *Superluminal Radio Sources* (Cambridge: Cambridge University Press).

Ultra-relativistic Pulsar Wind

Zhi-yun Li [1], Mitchell C. Begelman [1] and Tzihong Chiueh [2]

[1] Joint Institute for Laboratory Astrophysics, University of Colorado and
National Institute of Standards and Technology
[2] Institute of Physics and Astronomy, National Central University, Taiwan

Abstract: We show that cold, ultra-relativistic hydromagnetic winds can have a kinetic energy flux much larger than the Poynting flux. Even slight deviation of streamlines from a perfect radial geometry can move the fast magnetosonic point to some finite radii comparable to that of the light cylinder. Once the flow becomes supermagnetosonic, divergence of the streamlines can accelerate it via the "magnetic nozzle" effect, thereby converting Poyting flux to kinetic energy. We discuss our results in the content of pulsar winds.

1 Introduction

Shortly after the discovery of pulsars, it was proposed that a significant portion of the spindown energy could go into a relativistic wind propelled by the "magnetic slingshot" effect. By considering a relativistic ($\mu \gg 1$, μc^2: specific total energy) radial cold wind operating under this mechanisim, Michel (1969) showed that the Poynting flux always dominates the kinetic energy in the ratio $\mu : \mu^{1/3}$. However, models for the interior of the Crab nebula (Rees and Gunn 1974; Kennel and Coroniti 1984) require that the kinetic energy flux in the pulsar wind be several hundred times larger than the Poynting flux when the wind reaches the standing shock. This forced Kennel and Coroniti (1984) to conclude that the wind could therefore not be driven primarily by the magnetic slingshot effect, but rather must be driven thermally by the pressure of relativistically hot plasma.

Large ratio of kinetic energy to Poynting flux, however, is possible when the flow is collimated. In this case, slight deviation from a perfect radial wind geometry can move the fast magnetosonic point, which is always at infinity for a radial cold wind (Goldreich and Julian 1970), rapidly toward the light cylinder. Further acceleration of the flow beyond the fast point is obtained through a "magnetic nozzle" effect, provided that the streamlines diverge sufficiently.

2 Location of the Fast Magnetosonic Point

For a axisymmetric, steady-state, relativistic cold MHD wind, the conservation of the total specific energy μc^2 and angular momentum l, when combined with the flux frozen condition, yields an energy equation:

$$\left(\mu - \frac{x_A^2 - \mu x^2}{1 - x^2(1+\tau)}\right)^2 = 1 + (a\tau)^2 + \frac{x_A^4}{x^2}\left(1 - \frac{1 - \mu(x/x_A)^2}{1 - x^2(1+\tau)}\right)^2 \qquad (1)$$

where

$$x \equiv \frac{R\Omega}{c}, \quad x_A \equiv \sqrt{\frac{l\Omega}{c^2}}, \quad a \equiv \frac{B_p R^2}{kc^3/\Omega^2}, \quad \tau \equiv \frac{k^2}{4\pi\rho x^2}$$

and the mass flux per unit magnetic flux k and the angular velocity Ω are constant along a flux surface, while other quantities have their standard meanings. If the flow starts from x=0 at rest, then $x_A = \sqrt{\mu - 1}$.

The flow poloidal 4-speed $u_p(= a\tau)$ is completely determinated by the local value of the quantity a through eq. (1). In case of a radial wind, a is just the Michel's (1969) magnetization parameter σ, which does not change throughout the flow. For collimated flows, the deviation from a perfect radial wind is measured by a dimensionless parameter $\delta(x) = d \ln a/d \ln x$.

Making use of the fact that, at the fast point, two branches of root of eq. (1) cross, one can relate the critical energy μ_c, the deviation δ_f and the kinetic energy γ_f at the fast point to the location of the fast point x_f. The results are plotted in Fig. (1), from which we conclude:

(1). Small deviation δ can indeed move the fast point to a few light cylinder radii.
(2). Critical total specific energy μ_c is close to the value of a at the fast point a_f.
(3). The Poynting flux is still much larger than the kinetic energy up to the fast point, and further acceleration beyond the fast point is needed to have a kinetically dominated flow.

3 Acceleration Beyond the Fast Point

Well beyond the light cylinder ($x \gg 1$), the energy equation is simplified to:

$$\left(\frac{\tau}{1+\tau}\mu\right)^2 = 1 + (a\tau)^2. \qquad (1a)$$

If we denote the ratio of the Poynting flux to the kinetic energy by σ, then $\sigma = 1/\tau$. According to the above equation, an equipatition ($\sigma = 1$) is reached when $a \cong \mu/2 \cong a_f/2$. To get even smaller σ, a small $a(\ll \mu \gg 1)$ is required. In this limit, two branches of solution to eq. (1a) exist, namely:

$$\sigma \cong \frac{a}{\mu} \cong \frac{a}{a_f} \ll 1$$

and

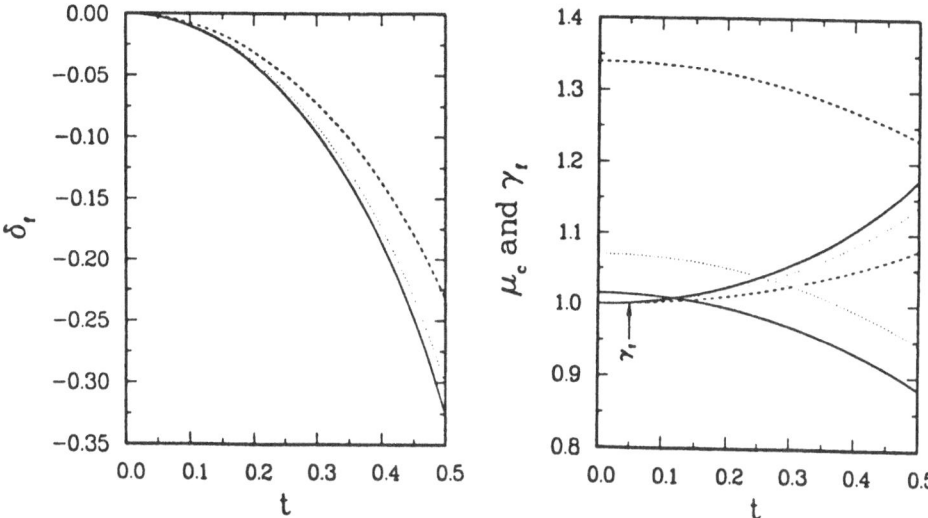

Fig. 1. (a) the amount of deviation from a perfect radial wind δ_f and (b) the critical specific energy μ_c (in units of a_f) and the specific kinetic energy γ_f (in units of $\mu_c^{1/3}$) are plotted against the position of the fast point $t \equiv 1/x_f$ for $_f = 10$ (dashed), 10^2 (dotted) and 10^3 (solid).

and

$$\sigma \cong \mu \gg 1$$

where the flow is supermagnetosonic in the former while submagnetosonic in the later. This behavior is analogous to hydrodynamic flow through a *de Laval* nozzle. Apparently, the solution with $\sigma \ll 1$ is what we want. Therefore, nearly all energy could be converted eventually into kinetic form provided that the field lines have such a geometry that the quantity a decreases significantly from its value at the fast point a_f to its asymptotic value a_∞ ($a_\infty/a_f \ll 1$) once the fast point is passed.

4 Asymptotic Field Structure

Studies on the asymptotic structure of hydromagnetic winds (Hayvearts and Norman 1989; Chiueh, Li and Begelman 1990) showed that it is possible to have flows with vanishingly small a_∞. In particular, there exists a family of paraboloidal flux surfaces that carries no net current asymptotically. Along these flux surfaces, the quantity a vanishes logarithmically. To illustrate this point, let us consider a general asymptotic expansion of the paraboloidal field line configuration as

$$z = f(\psi)R^{\alpha(\psi)} \tag{2}$$

where ψ is the flux function defined from $\mathbf{B}_p = \nabla\psi \times \hat{e}_\phi/R$ and $\alpha(\psi) > 1$. It follows that

$$a \propto B_p R^2 \cong \left| R \frac{\partial \psi}{\partial R} \right| \cong \frac{\alpha}{|\alpha' \ln R + (\ln f)'|} \propto \frac{1}{\ln R}$$

provided that $\alpha'(\psi) \neq 0$.

It is not guaranteed, however, that a flow with the asymptotic field structure given by eq.(2) can pass smoothly through the critical points and connect to the flow base. But if it does, then a kinetically dominated cold wind is obtained.

5 Application to the Crab Pulsar Wind and Other Objects

The ratio of the Poynting flux to the kinetic energy σ in the Crab pulsar wind is estimated (Kennel and Coroniti 1984) to be 0.003 when it is terminated by a strong standing shock at a distance of about 3×10^{17}cm from the pulsar. Such a small σ cannot be due to the logarithmic effect disscussed above alone, since it can only yield a factor of about 20 at most, assuming the fast point is not far away from the light cylinder of 1.6×10^8cm. It is possible, however, to envision an acceleration region right after the fast point where a can decrease faster than $1/\ln R$, so that a/a_f could be as low as required.

In summary, although the conditions neccessary to obtain $\sigma \sim 0.003$ or less might be extreme, a wind with comparable energy in the kinetic and electromagnetic form is certainly obtainable, in contrast to a cold radial wind.

References

Chiueh, T., Li, Z. and Begelman, M. C. 1990, *Ap. J. (Letters)*, (submitted).
Goldreich, P., and Julian, W. H. 1970, *Ap. J.*, **160**, 971.
Hayvaerts, J., and Norman, C. A. 1989, *Ap. J*, **347**,1055.
Kennel, C. F. and Coroniti, F. V. 1984, *Ap. J.*, **283**, 694.
Michel, F. C. 1969, *Ap. J.*, **158**, 727.
Rees, M. J. and Gunn, J. E. 1974, *M.N.R.A.S.*, **167**, 1.

A Model of Pulsed Gamma-Radiation from the X-Ray Binary Hercules X-1/HZ Herculis

F. A. Aharonian and A. M. Atoyan

Yerevan Physics Institute
Alikhanian Brothers St. 2, 375036 Yerevan, Armenia, USSR

Abstract: A model of pulsed VHE and UHE γ-radiation from X-ray binaries is proposed. The model implies that the γ-rays are due to the bombardment of a cloud ejected from the companion normal star by the relativistic proton beam stationarily accelerated by the pulsar. In the framework of this model, all the peculiarities of the γ-radiation observed Hercules X-1/HZ Herculis are naturally explained. Namely, a) the γ-ray pulsation frequency shift with the respect to the X-ray frequency; b) episodic nature of γ-ray events with typical burst duration $\Delta t \lesssim 1$ hour; c) the absence of any correlation between the γ-ray events and the orbital phase of the binary; d) the observation of γ-ray events in the phase of the deep eclipse of the pulsar. The expected γ-ray spectra in a wide range of energy, $100 \text{ MeV} < E < 1 \text{ PeV}$, as well as the possibilities of experimental verification of the model are discussed.

Gamma-ray Astronomy Above 100 GeV

Giuseppe Vacanti

Service d'Astrophysique, Centre d'Etudes Nucléaires de Saclay
91191 Gif-sur-Yvette Cedex
France
SPAN: 32779::vacanti

Abstract: The present status of VHE and UHE γ-ray astronomy is reviewed. Some emphasis is given to experimental techniques to illustrate current problems and attainable sensitivities. An all-source catalog is presented; five sources are then reviewed in some detail.

1 Introduction

Gamma rays trace high energy processes in the Universe. Travelling almost unperturbed through the Galaxy they point directly to the location of their sources and, coupled to observations in lower energy bands, help unveil the mechanisms at work in *Cosmic Particle Accelerators.*

In recent years the field of ground based γ-ray astronomy ($E_\gamma > 100\,\text{GeV}$) has attracted increasing interest, perhaps partly due to the fact that the satellite window ($E_\gamma \approx 100\,\text{MeV}$) has been inaccessible for almost 10 years now. The growth is manifest in the number of experiments now in operation around the world (> 30) and in the number of papers that can be found in the proceedings of the International Cosmic Ray Conferences (more than 100 for the last Conference versus ≈ 10 a decade ago). Radio pulsars, X-ray binary systems, supernova remnants, millisecond pulsars, cataclismic variables: these are the types of astrophysical objects under investigation by the ground based community. More than 20 are now claimed to be sources of γ rays in the Very High Energy and Ultra High Energy ranges. Some of the evidence gathered by experimenters on some of these objects will be reviewed in the following, preceeded by an outline of the experimental techniques used[1].

---

[1] More complete reviews are given by Weekes 1988; Nagle, Gaisser, and Protheroe 1988; Fegan 1990.

2 Detection Techniques

The Struggle for Sensitivity

If there is anything like a typical TeV/PeV γ-ray source, we can expect it to emit $10^{-10} - 10^{-11}\,\gamma\,\mathrm{cm}^{-2}\,\mathrm{s}^{-1}$ above 1 TeV and $10^{-13} - 10^{-14}\,\gamma\,\mathrm{cm}^{-2}\,\mathrm{s}^{-1}$ above 1 PeV. This assumes an integral photon spectrum proportional to E^{-1}; such a behaviour is what observations of Cygnus X–3 and Vela X–1 seem to suggest, but given the present uncertainties this may be an optimistic assumption. A detector located just outside the Earth's atmosphere and with an effective collection area of $1\,\mathrm{m}^2$ (to be compared to $0.1\,\mathrm{m}^2$ at 100 MeV for the soon-to-be-launched EGRET on GRO) would detect a few tens of photons with energy greater than 1 TeV in one year of uninterrupted observation; proportionally less above 1 PeV. No reasonable astronomy could be developed in these conditions. Fortunately, ground based techniques overcome these limitations: entering the atmosphere a photon with $E_\gamma \geq 100\,\mathrm{GeV}$ generates an Extensive Air Shower (EAS) that, by the time it becomes available for detection, forms a pancake-like front (composed mainly of electrons, positrons, and other photons) a few meters thick and with a lateral extent in excess of 100 m. The size of this front of secondaries defines the collection area of the detector which can therefore be of the order of a few hectars. The huge collection area gives ground based techniques the potential to detect low fluxes in a reasonable amount of time once the background is to some extent reduced. In fact EAS's induced by cosmic rays look very much alike the γ ones and they are much more numerous ($\frac{N_\gamma}{N_{cr}} \leq 10^{-4}$ at the sources; i.e. Gaisser 1990). Effective background discrimination becomes a *sine-qua-non* for the development of the field. It can be achieved by direct recognition of the photonic nature of the showers or by improving the angular resolution of the detectors.

2.1 Very High Energy: 100 GeV – 100 TeV

The particle component in a shower generated by a primary with energy in the VHE range is almost completely absorbed before it reaches an altitude accessible for ground detection. This is not the case for the Cherenkov light the particles emit while travelling in the atmosphere at relativistic speeds. Conventional optical techniques can be used to detect the Cherenkov light flashes induced by cosmic primaries. The Atmospheric Cherenkov Technique consists then in detecting optical flashes induced by EAS's in the atmosphere against the night-sky background light. In spite of the signal's low intensity, this is possible because a Cherenkov light flash lasts less than 10 ns: no other atmospheric phenomenon can produce impulses on this time scale.

The do-it-yourselfer's kit for an atmospheric Cherenkov detector consists of a parabolic mirror about 2 meters in diameter, a fast photomultiplier and some pulse counting electronics. The parameters of the detector can be adjusted so that it have a field of view of $2° - 3°$ fwhm, an effective energy threshold of a few TeV and a collection area of $\approx 5 \times 10^4\,\mathrm{m}^2$. As remarked before, such a detector is marred with a low signal-to-noise ratio so that very long exposures are necessary to achieve statistically significant detections. Simple ways to reduce the sensitivity of the detector to cosmic rays (for example by decreasing the field of view) always result in a reduced sensitivity to γ's as well (this happens because the collection area for γ's is reduced in the process), so that in its simplicity and potential the basic atmospheric Cherenkov detector is hardly improvable.

Two approaches can be followed in order to increase the sensitivity at the expense of simplicity:

  a. Determine where in the field of view the Cherenkov triggers are coming from (i.e. improve the angular resolution without decreasing the field of view and the collection area).
  b. Individuate the nature of the primary by recognizing some characteristics of the Cherenkov light front.

Almost all detectors currently in operation exploit (a), or (b), or both. Two of them will be briefly described in the following.

2.1.1 Asgat

Located on the French side of the Southern Pyreneés at an altitude of 1700 m, the Asgat (**AS**tronomie **G**amma **A T**hémis) telescope consists of seven parabolic mirrors, each 7 meters in diameter, arranged in an hexagonal grid ($\approx$ 40 m on a side) with one mirror in the centre (Goret et al. 1988; Basiuk et al. 1989). Signals seen by at least four of its detectors are fast timed to individuate the incidence direction of the Cherenkov light front (hence the generating primary's). The principle of the technique is illustrated in Figure 1: the angular resolution achievable is $\delta\alpha = \frac{c}{d} \cdot \frac{1}{\cos\alpha} \cdot \delta t$, where c is the speed of light. With nanosecond time resolution and letting $d = 50$ m $\delta\alpha \approx 0.3°$ at the zenith (in the case of Asgat, for an event seen by at least 4 detectors $\delta\alpha \approx 0.2°$ rms).

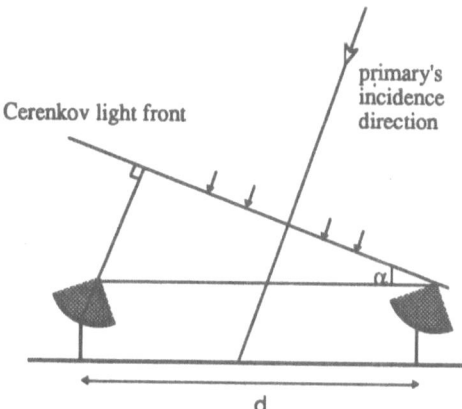

Fig. 1. Principle of the fast timing technique (courtesy of Basiuk).

The underlying assumption in the above estimates is that the Cherenkov light front can be approximated by a plane wavefront. In reality the front is roughly spherical with a radius of curvature of several kilometers: the center of the sphere coincides approximately with the location of the maximum development of the shower. Sampling the front at a few points spread at the most over a few tens of meters will not allow to recognise the curvature: this introduces an error in the accuracy with which the incidence direction of the primary can be reconstructed. The magnitude of this parallax effect can be estimated by Monte Carlo simulations: in the case of Asgat it is found to be $\approx 0.5° - 0.6°$ rms. In its present configuration the detector operates at an energy of 400 GeV and can detect at the 3σ level a flux $\approx 10^{-10}$ cm^{-2} s^{-1} in 30 hours on source (at least the same amount should be spent on a suitable control region).

2.1.2 The Whipple Imaging Detector

A different approach consists in attempting to recognise the nature of the primary generating each EAS. The starting point is to ask what are the differences between a photonic and a hadronic EAS, and whether any of these differences are somehow preserved after the shower's homogenising journey through the atmosphere. Figure 2 shows simulations of a γ-induced and a proton-induced EAS: only the particle components are shown. The photonic shower develops higher in the atmosphere and, being basically only electromagnetic in nature, is more compact and beamed forward along the incidence direction of the primary photon. This is not the case for the protonic shower due to transversal momenta generated in nuclear interactions. If the Cherenkov light emitted by each particle in the shower is followed through the atmosphere to ground level, some of these characteristics are still preserved: the γ-induced Cherenkov light pool appears to be more regular and compact (Figure 3). When the Cherenkov photons are propagated to the focal plane of a light collector it can be shown that γ-induced EAS have a smaller angular size and tend to have an approximately elliptical shape with the major axis pointing towards the position of the source (Hillas and Patterson 1987). Differentiation between the two types of shower can be achieved with a detector able to image the Cherenkov light pools on the ground.

This imaging approach is being pursued by the Whipple Collaboration. The Collaboration operates the 10 meter light collector on Mt. Hopkins, Arizona (2320 m asl). At the focal plane a 109-pixel (photomultipliers) camera is used to digitise Cherenkov events (Cawley et al. 1990). Off line, parameters characterizing shape and orientation of the images in the focal plane are computed and used to classify the events in two categories: probable photon and probable hadron. One parameter, *Azimuthal-Width* (Figure 4), has proven to be most effective. As suggested by simulations (Hillas 1985) and confirmed by the Collaboration's reports on the Crab Nebula (Table 1), 97% of the hadronic background can be rejected while retaining 60% of the γ-ray signal by discriminating events on the base of their *Azimuthal-Width*. In the present configuration the Whipple telescope, operating at an energy of 400 GeV, can detect a minimum flux of $7 \times 10^{-12} \gamma$ cm^{-2} s^{-1} at the 3σ level in 70 hours of observation (of which half spent on a control region).

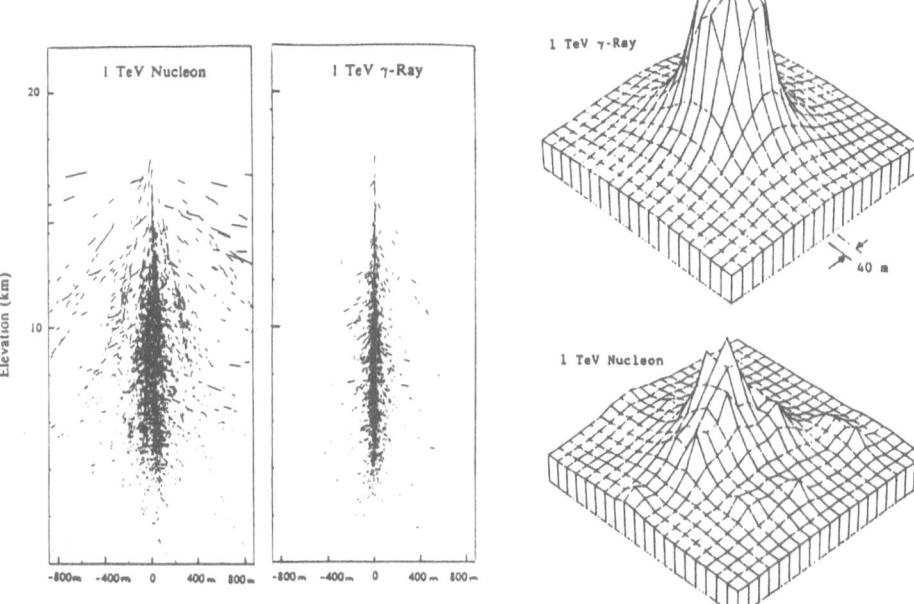

Fig. 2. Longitudinal development of Extensive Air Showers. Only the particle component is shown.

Fig. 3. Cherenkov light pools generated by the showers of Figure 2.

Table 1. Twenty years of Crab Nebula observations with the Whipple telescope.

Epoch	Time ON source(h)	Imaging?	σ	Reference
1969-70	150	no	3.1	Fazio et al. 1972
1986-88	80	yes: 37 pixels	8.9	Weekes et al. 1989
1988-89	32	yes: 109 pixels	15.3	Lang et al. 1990

The imaging technique is presently being carried a step forward with the construction of a second imaging telescope that will be placed at a distance of about 120 m from the present one. The symbiotic use of two independent imaging detectors will gain a further factor 3 in sensitivity. This project, named GRANITE (Gamma Ray Astronomy New Imaging TElescope), should be in operation by Autumn 1991 (Akerlof et al. 1990).

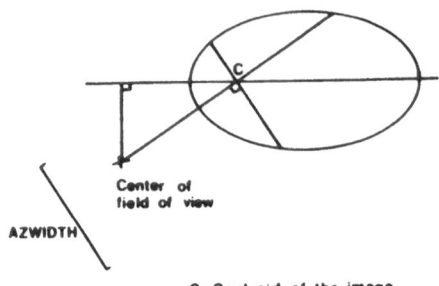

Fig. 4. Definition of *Azimuthal-Width*.

2.2 Ultra High Energy: 100 TeV – 100 PeV

At energies above 100 TeV the particle component of an EAS can be detected directly on the ground via an array of particle detectors. The direction of the shower is reconstructed by fast timing the response of the single detectors. The electron density distribution is also sampled and iterative fitting procedures are used to estimate the number of electrons in the shower (hence its energy), its impact parameter and age s (see below). Performance of each array (energy threshold, angular resolution etc.) depends on many factors (total area covered, active detecting area, spacing of the detectors etc.). As a reference, the Chicago Air Shower Array, now being completed in Utah, will consist of 1064 detectors (of which about 600 are already in place) arranged in a square grid 480 m on a side. Scheduled to be completed by the Summer 1991, it will have an angular resolution of $\approx 0.6°$ rms, an effective energy of 100 TeV and is expected to detect a minimum flux of 10^{-13} $cm^{-2}\,s^{-1}$ at the 5σ level in one year (Gibbs et al. 1988).

One can again ask whether the nature of the primary can be identified by looking at some parameters describing the secondaries. Two approaches are followed by the practitioners of the field.

a. *Muon content* Owing to the low photon hadron-production cross section, γ-induced showers are expected to contain less than 10% the number of muons seen in hadronic showers. This is confirmed by detailed Monte Carlo calculations of shower development in the atmosphere (e.g., Stanev, Gaisser and Halzen 1985). Tagging of μ-poor showers should in principle discriminate γ-induced showers from the cosmic-ray-induced background (see paragraphs 3.2 and 3.3 for some experimental results on this point). The CASA array will operate at the same site of the MIchigan Muon Array. This consists of a total of $\approx 1300\,m^2$ of moun detectors buried beneath 3 m of earth. The detectors are arranged in 8 clusters under the CASA array (Figure 5). The CASA-MIA ensemble should reject EAS of cosmic ray origin with high efficiency ($\approx 99\%$), bringing the minimum flux detectable in one year at the level of $\approx 10^{-14} cm^{-2}s^{-1}$ (Gibbs et al. 1988).

b. *Age of the showers* The age of a shower s is a parameter used to describe its development stage. A shower is given an age $s = 1$ when it is at its maximum development; $s < 1(> 1)$ for a younger (older) shower, that is to say one sampled before (after) it has completely developed. The age of the shower is estimated by fitting a suitable lateral distribution function to the sampled electron density distribution. The reasoning behind the use of the age parameter as a discriminator rests on the consideration that because γ showers tend to get initiated higher in the atmosphere, at a given altitude they will look on the average older than their hadronic counterparts. Some of the UHE results found in the litterature seem to confirm this line of thinking (see paragraphs 3.2 and 3.4 below). However Hillas (1987) has shown that the age of a shower is a very poor discriminator and that, if anything, γ showers should appear younger, not older, than the background, so that at this time there appears to be no physical footing for the use of this parameter to select photonic showers. Perhaps the present situation is best summarised by the words of Ciampa et al. (1989): *it is not clear why this* [selecting old EAS's] *might preferentially select γ-ray induced showers, but in a number of cases it appears to have enhanced a possible signal.*

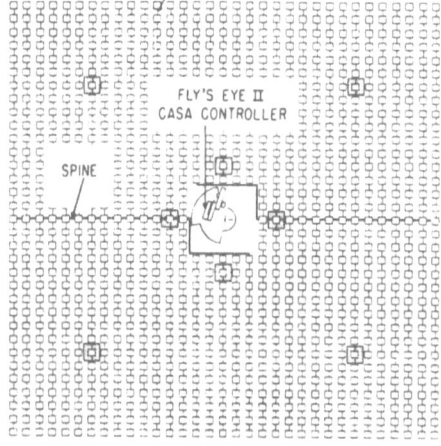

Fig. 5. Lay-out of the CASA-MIA detectors (Gibbs et al. 1988).

3 Source Catalog

A review of the VHE/UHE source catalog cannot but begin with a few words about the techniques of data analysis and the statistics used to estimate the significance of possible signals. There are two basic ways to look for a signal: DC, when a count excess is sought for in the direction of a particular object, and AC, when the data base is searched for some periodic signature identifying a source. In both cases the signals are weak, may be sporadic and the data bases are background dominated. As a consequence it is very important to know the statistical properties of the background in order to properly evaluate the significance of a putative signal. Slight variations in the expected Poisson behaviour, if overlooked, could severely bias the results of a given experiment. Another aspect to take into account in the evaluation of the significance of a result is the number of degrees of freedom, explicit and implicit, introduced into the analysis procedure. This is relatively straightforward to do when dealing with DC searches, but it becomes more subtle when AC analyses are carried out. Some of the problems to deal with are: proper estimation of the oversampling factor when searching for a signal over a wide range of frequencies (de Jager 1987); evaluation of the distorsions produced in the Fourier power spectrum by the analysis of gapped data (Lewis, Lamb, and Biller 1990); compatibility of AC and DC signals (de Jager, Raubenheimer, and Swanepoel 1988; Lewis 1989); meaning of the concept of independent Fourier spacing when dealing with unevenly sampled time series (Koen 1990). It is fair to say that only a few years ago practitioners of the field did not even know that some of these problems were there. Some statistical work has come out of the field but we are far from understanding all the implications of these issues: this is an extra hurdle to overcome in *the Struggle for Sensitivity*.

The following two examples should help clarify some of the problems.

First Example: in 1985 the Potchefstroom group (Raubenheimer et al. 1986) provided evidence for TeV γ-ray emission from PSR 1802-23 (112.5 ms). This report was very interesting in that this pulsar lies inside the error box of the COS-B source 2CG006-00: the effect was estimated at the 99.8% confidence level. More data were collected in 1986, 1987, and 1988 to seek confirmation of the result; at the same time a reanalysis of the 1985 data was undertaken using a Monte Carlo method to better evaluate the significance of the original claim. As a result the claim was withdrawn as not significant (Nel et al. 1990). In the meantime reanalyses of the COS-B data base in view of new information about the distribution of matter in the Galaxy have shown that 2CG006-00 may not be a γ-ray source, but only an effect due to the interstellar gas (Mayer-Hasselwander and Simpson 1990).

Second Example: the Whipple group reported TeV emission from the X-ray binary 4U0115+63 (Lamb et al. 1987), source already claimed by Chadwick et al. (1985) and possibly by the Crimean Astrophysical Observatory in 1972 (in Lamb and Weekes 1986). Coherently linking in phase data from 3 consecutive nights a significant peak in the Fourier spectrum was seen after searching on both the neutron star period and its derivative. Recent work carried out on the distorsions produced in the Fourier spectrum by searching gapped data in parameter space (Lewis, Lamb, and Biller 1990) leads to the conclusion that the original claim was unwarranted: the upper limits derived are in contrast with the flux values reported by other observers (Macomb et al. 1991).

With these caveats in mind, we can move on to browse through the VHE/UHE source catalog (Table 2). In compiling such a catalog, it is difficult to decide on the degree of belief to attach to a given source. Many objects score only one published paper; others boast many claims who cannot be reconciled. At times a claim is made by an experiment still in its commissioning phase who then fails to gather confirming evidence once it has supposedly become more reliable (time variablity of the source can almost always be invoked in these cases ...). Many claims are not confirmed by an independent group: this should probably be taken into account but it is not clear how. Also, the community is slow in applying retroactively new statistical tools to already analysed data bases, so that marginal detections neither get strengthened nor are definitively rejected (but always lurk in the dark embarassing the doubtful Reviewer!!).

Having said that, Table 2 lists sources claimed by at least one refereed paper. A section on would-like-to-be sources is also included: it lists claims made in conference proceedings but, to this Reviewer's best knowledge, still lacking the *imprimatur* of a referee.

In the following more attention will be given to five sources. The choice, forcibly limited, is none the less believed to represent a good sample of the objects and problems currently debated in the ground based community.

3.1 Crab Nebula and Pulsar

This system has attracted the attention of the ground based community since the late 60's. In the VHE regime the Nebula was first observed by the Smithsonian Astrophysical Observatory (Fazio et al. 1972) using the 10 meter optical reflector on Mt. Hopkins, Arizona. Over the years many groups have confirmed its existence at comparable flux levels, a remarkable fact in itself. Some of the recent reports are summarised in Figure 6. The signal, albeit weak, appears to have been constant at least over the past 20 years: this makes the Nebula a standard candle for detectors who have access to it. Also shown in Figure 6 is the energy spectrum derived by the analysis of the Whipple 1988-90 data base (Weekes 1990). This is the first time some spectral information is obtained at these energies. The spectrum can be written as $9.8 \times 10^{-11} \cdot \left(\frac{E_{TeV}}{0.4 \cdot TeV}\right)^{-2.4 \pm 0.3} \gamma \, cm^{-2} s^{-1} \, TeV^{-1}$. Although the uncertainties are still big, this result puts some constraints on models proposed to explain the nebular emission at this energy.

At higher energies the situation is not as clear cut. There are claims from the Lodz (Dzikowski et al. 1983) and Tien-Shan (Kirov et al. 1985) arrays for UHE emission from the Nebula (the flux levels are comparable, $2 - 3 \times 10^{-13} cm^{-2} s^{-1}$, however the two arrays have thresholds that differ by a factor of 30). On the other side, data from the Haverah Park Array, with better angular resolution, show no excess whatsoever (Watson 1985). The limit derived is about two orders of magnitude lower than the Lodz limit at the same energy. More recently, the Los Alamos group have found a 2.5σ excess in their 1986-90 data base from the direction of the Crab Nebula (Kwok 1990). They use this to set an upper limit to the continuous flux above $50 \, TeV$ of $6.4 \times 10^{-14} cm^{-2} s^{-1}$ (90 % confidence level). This result is compatible with the extrapolation of the spectrum derived by the Whipple group.

The situation is confused as far as the emission from the pulsar is concerned. There can be little doubt that PSR 0531+21 accelerates electrons up to $100 \, TeV$ and is the ultimate power source for the Nebula. Whether in the process TeV γ rays can be produced and escape the intense magnetic field in the vicinity of the neutron star depends on a favourable configuration of the system. Over the past 20

Table 2. All-source catalog

Name	Detections	Periodiciy	Energy	Flux $(cm^{-2} s^{-1})$	Luminosity (erg/s)
Supernova **Remnants**					
Crab Nebula	3	-	VHE	5×10^{-11}	$2 \times 10^{+34}$
	1		UHE	1×10^{-13}	$2 \times 10^{+35}$
Pulsars					
PSR 0355+54	1	156 ms	VHE	8×10^{-12}	$3 \times 10^{+34}$
PSR 0531+21 (Crab)	6	33 ms	VHE	4×10^{-12}	$6 \times 10^{+33}$
	(1)	33 ms	UHE	4×10^{-13}	-
PSR 0833-45 (Vela)	2	89 ms	VHE	3×10^{-12}	$3 \times 10^{+34}$
PSR 1509-08	2	150 ms	VHE	2×10^{-11}	$5 \times 10^{+34}$
PSR 1953+29 (2CG065 ?)	1	6 ms	VHE	3×10^{-11}	$3 \times 10^{+35}$
PSR 1937+21	(1)	1.5 ms	VHE	2×10^{-11}	$2 \times 10^{+35}$
X-ray **Binaries**					
Her X-1	3	1.24 s	VHE	3×10^{-11}	$3 \times 10^{+35}$
	1	1.24 s	UHE	3×10^{-12}	$2 \times 10^{+37}$
Cygnus X-3	10	4.8 h & 12.6 ms	VHE	10^{-11} 10^{-13}	$3 \times 10^{+36}$
	5	4.8 h	UHE	2×10^{-14}	$6 \times 10^{+36}$
Vela X-1	1	283 s	VHE	2×10^{-11}	2×10^{-34}
	1	9 d	UHE	10^{-14}	2×10^{-34}
Sco X-1	1	-	VHE	10^{-10}	$2 \times 10^{+34}$
2A 18822-37.1	1	.2 d	UHE	10^{-15}	-
LMC X-4	1	1.4 d	UHE	5×10^{-15}	-
Cen X-3	(2)	4.8 s	VHE	6×10^{-10}	$6 \times 10^{+36}$
4U0115+63	2	3.6 s	VHE	7×10^{-11}	$6 \times 10^{+35}$
Extragalactic					
Cen A	1	-	VHE	4×10^{-12}	$3 \times 10^{+40}$
M 31	1	-	VHE	2×10^{-10}	$4 \times 10^{+40}$
Proposed					
E0021.8-7221 (Tucane 47)	-	120 s	VHE	-	$2 \times 10^{+35}$
1E2259+586	(1)	7 s	VHE	2×10^{-10}	-
PSR 0950+08	1	-	VHE	1×10^{-11}	-
AE AQR	-	33 s	VHE	-	-
PSR 1855+09	-	-	VHE	-	-
PSR 1957+20	-	-	VHE	-	-
Geminga (2CG195-4)	-	60 s?	VHE	-	-

years there have been many reports claiming to see the 33 ms signature (reviewed in Weekes 1988). Often the claims are not very convincing in themselves. Many light curve profiles have been reported that do not agree on the phase of the emission, the number and widths of peaks in the light curve. Also, some groups claim a steady pulsed signal, while others have shown evidence for burst-like activity on a

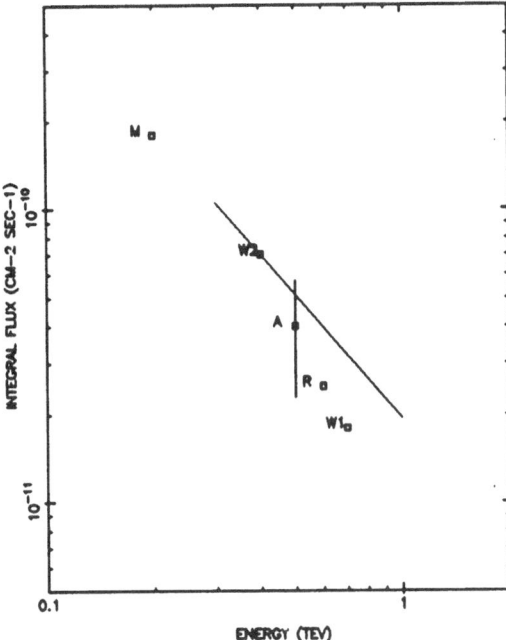

Fig. 6. Recent detections of the Crab Nebula in the VHE range. W1: Weekes et al. 1989; M: Akerlof et al. 1990; W2: Lang et al. 1990; R: Turner et al. 1990; A: Basiuk et al. 1991.

15 minute time scale. Among all, the report by the Durham group (Dowthwaite et al. 1984a) stands out for its significance. About 100 hours of data collected in 1982-83 show a clear signal when folded with the pulsar radio ephemeris. The chance probability of such an effect was estimated in $\approx 10^{-5}$, and a time averaged pulsed flux of $(7.9 \pm 1.8) \times 10^{-12}\,\mathrm{cm}^{-2}\mathrm{s}^{-1}$ was inferred above 1 TeV. Attempts have been made by the Tata (Bath et al. 1990), Michigan (Akerlof et al. 1990), and Whipple (Weekes et al. 1989; Lang et al. 1990) groups to confirm the result without success. The limits reported make understanding of the Durham claim difficult (Figure 7).

In the UHE range the Tata group report to have found the pulsar in data collected during 1984-87 with the Ooty air shower array (Gupta et al. 1990a). The emission is seen in coincidence with the radio interpulse. The flux above 200 TeV is estimated to be $(4.1 \pm 0.8) \times 10^{-13}\,\mathrm{cm}^{-2}\mathrm{s}^{-1}$ (99.8% confidence level).

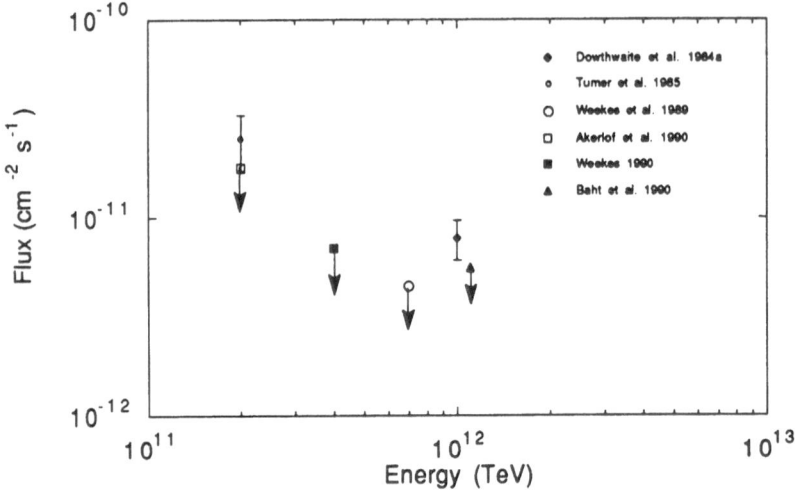

Fig. 7. Recent VHE reports on PSR 0531+21.

3.2 Cygnus X-3

Probably one of the most looked at objects in the sky, more than 20 year after its discovery in the X-ray band Cygnus X-3 still defies our understanding. The nature of the system is unknown: the 4.8 h modulation in the X-ray emission is usually ascribed to the orbital motion of a compact object (neutron star) around a companion. The source lies very close to the galactic plane in a region dense in interstellar dust that prevents any optical identification. Measurements in the radio and X-ray bands yield only lower limits to its distance (> 8 Kpc). Cygnus X-3 is subject to periods of very intense activity with strong bursts in the radio band, where it becomes one of the brightest sources in the sky for periods ranging from a few days to a few weeks.

Like the Crab, Cygnus X-3 has long been on the target board of γ-ray astronomers, its first detection in the VHE range dating 1972 (Vladimirsky, Stepanian and Fomin 1973). Over the past 18 years there have been many claims and at least as many refutations concerning its existence and phenomenology.

After the initial reports, Cygnus X-3 was elevated to primary topic of the field by the Kiel group observation in the UHE range (Samorsky and Stamm, 1983a). A 4.4σ excess from the direction of Cygnus X-3 was seen in data collected between 1976 and 1980 (Figure 8). The excess was seen only in showers with $s > 1.1$. The flux above 2×10^{15} eV was estimated in $(7.4 \pm 3.2) \times 10^{-14}$ cm^{-2} s^{-1}. The signal also showed to be modulated with the 4.8 h periodicity. A puzzle was brought about by this detection: analysis of the muon content showed that the signal showers had 75% of the muons found in background events (Samorsky and Stamm 1983b). This is inexplicable if the primaries are photons, as it was supposed on the basis of the age selection (e.g. Stanev, Gaisser and Halzen 1985).

The literature abounds with claims and counter claims about this object: no attempt will be made to review them here. For critical reviews of the experimental

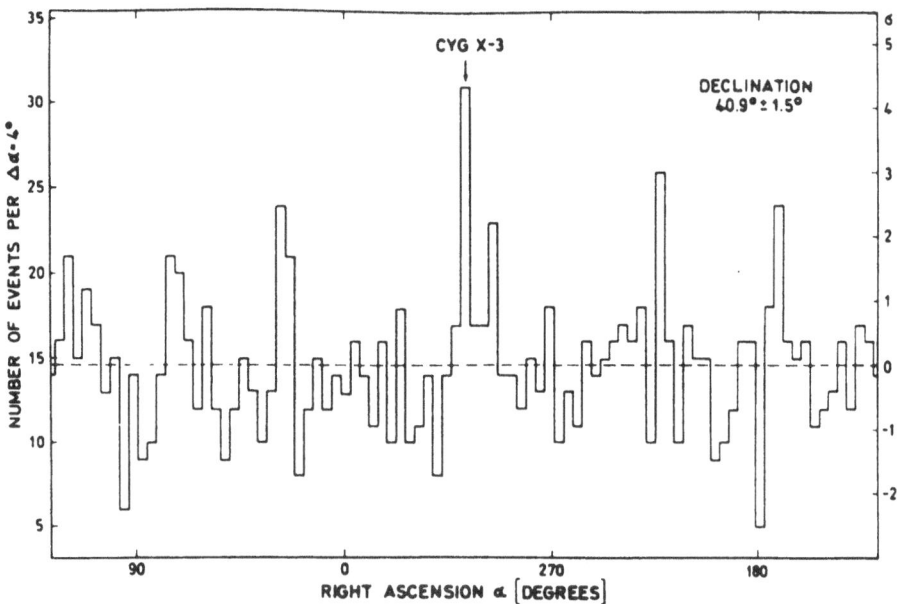

Fig. 8. Kiel group detection of Cygnus X-3 (Samorsky and Stamm 1983a).

situation concerning Cygnus X-3 see Chardin and Gerbier (1989), and Bonnet-Bidaud and Chardin (1989).

Recent reports have not improved the situation. The Fly's Eye group (Cassiday et al. 1989) reported a flux of $(2.0 \pm 0.6) \times 10^{-17} \mathrm{cm}^{-2} \mathrm{s}^{-1}$ $(E \geq 5 \times 10^{17} \mathrm{eV})$ from Cygnus X-3 inferred from data collected between 1981 and 1988: the confidence level of the detection is estimated in 99.93%. There is a weak 4.8h modulation.

Confirmation of this result seems to come from the Akeno air shower array. Teshima et al. (1990) analysed data collected in 1984-89 and found a flux of $(1.8 \pm 0.7) \times 10^{-17} \mathrm{cm}^{-2} \mathrm{s}^{-1}$ at the same energy as the Fly's Eye. The most significant bin in the sky map constructed with the Akeno data stands 3.5σ above the background, however it is displaced by about 2° from the position of Cygnus X-3. The authors estimate the probability of this happening by chance to be 10% and quote a 99.9% confidence level for their detection.

In the best tradition of the field, when two results seem to agree, a third one comes along to obfuscate the picture. The analysis of 13 years (1974-1987) of data collected on Cygnus X-3 by the Haverah Park array does not sustain the claims made by the Fly's Eye and Akeno groups. Lawrence, Prosser and Watson (1989) quote an upper limit of $4 \times 10^{-18} \mathrm{hadrons\,cm}^{-2} \mathrm{s}^{-1}$, always at an energy of $5 \times 10^{17} \mathrm{eV}$. The limit is $8 \times 10^{-18} \mathrm{cm}^{-2} \mathrm{s}^{-1}$ if the primaries are photons.

At VHE energies the attention has been concentrated on the detection of the millisecond pulsar expected to exist in the Cygnus X-3 system. Evidence for the its existence (spin period $\approx 12.6\,\mathrm{ms}$) has been provided at different times by the Durham group (Chadwick et al. 1985; Brazier et al. 1990). The pulsar is visible only at a well defined phase of the 4.8 h cycle ($\phi_{\mathrm{orb}} \approx 0.625$); also, it appears to be active only 10-20% of the time. The Durham group were able to determine a period derivative: the pulsar slows down at a rate of $3 \times 10^{-14}\,\mathrm{s\,s^{-1}}$. Their data suggest that above 1 TeV the time averaged flux is about $5 \times 10^{-12}\,\mathrm{cm^{-2}\,s^{-1}}$ with a flux around the maximum X-ray emission ($\phi_{\mathrm{orb}} \approx 0.625$) of the order of $10^{-10}\,\mathrm{cm^{-2}\,s^{-1}}$. Since it was first reported, three different groups have unsuccessfully attempted to confirm the existence of this pulsar (Fegan et al. 1989; Resvanis et al. 1987; Baht, Ramana Murthy and Visvanath 1988). Recently, possible confirming evidence on the existence of a millisecond pulsar in the system has come from Gregory et al. (1990). Using the atmospheric Cherenkov telescope they operate at Woomera, South Australia, they report on the detection of a millisecond pulsar in the Cygnus X-3 system. Data were collected on 5 nights during August-September, 1989, about 2 months after a radio burst. The period found, 12.5953 ms, differs from the Durham prediction (12.5962 ms) by twice the quoted uncertainty and comes about at a different orbital phase ($\phi_{\mathrm{orb}} = 0.564$, or 16 minutes too early).

3.3 Hercules X-1

The nature of the low mass X-ray binary system Her X-1 is very well understood. In the X-ray band three distinct periodicities are recognised. The neutron star spins with a period of 1.24 s and completes one orbit around the barycenter in 1.7 d. The X-ray emission is also modulated with a 35 d periodicity whit the source being ON over two intervals for a total of about 11 days. This modulation is usually ascribed to the precession of the accretion disk. Cyclotron absorption features in the X-ray spectrum allow to estimate the magnetic field of the neutron star to be $3 \times 10^{12}\,\mathrm{G}$ (Mihara et al. 1990). The system is also identified in the visible and infrared bands where it is known to episodically emit radiation at periods slightly different (0.1%) than that of the neutron star (Middletich et al. 1985). Long periods of quiescence characterise the system in optical and X-ray bands.

In the VHE range Hercules X-1 was discovered by the Durham group who observed a large increase in counting rate lasting $\approx 3\,\mathrm{minutes}$ (Dowthwaite et al. 1984b). The excess events were attributed to Her X-1 by recognising the 1.24 s pulsation. At UHE energies the first report came from the Fly's Eye group (Baltrusaitis et al. 1985). It has to be remarked that the Cherenkov telescopes of the Durham group, then located at the same site as the Fly's Eye, were simultaneously observing Her X-1 and did not report any sign of activity (Chadwick et al. 1985): a very flat energy spectrum is needed to explain the discrepancy.

Since 1984 several groups have reported episodes of VHE and UHE emission lasting between a few and 90 minutes. They are summarised in Figure 9 vis á vis the 1.7 d and 35 d cycles.

The year 1986 was apparently one of great VHE/UHE activity in the Her X-1 system: five groups have reported episodes of increased activity (Table 3).

As we shall see, four of the five groups were able to perform a timing analysis on their data and the results show striking similarities. Taken at face value this

Table 3. Reports on Hercules X-1 in 1986.

Group	Date	Energy (TeV)	Flux (cm^{-2} s^{-1})	Luminosity (erg/s)	Phase 1.24 s	Phase 1.7 d	Phase 35 d
Tata	April 11	.4	2x10^{-8}	2x10^{+37}	-	0.19	0.42
Haleakala	May 13	.4	3x10^{-9}	-	0.7	0.81	0.22
Whipple	June 11	.6	2x10^{-10}	-	0.7	0.7	0.04
Ooty	July 1	100	3x10^{-11}	4x10^{+37}	0.3	0.65	0.65
Los Alamos	July 24	200	2x10^{-11}	-	?	0.85	0.23
Ooty	August 8-9	100	3x10^{-11}	4x10^{+37}	0.3	0.65	0.65
Ooty	November 21	same					

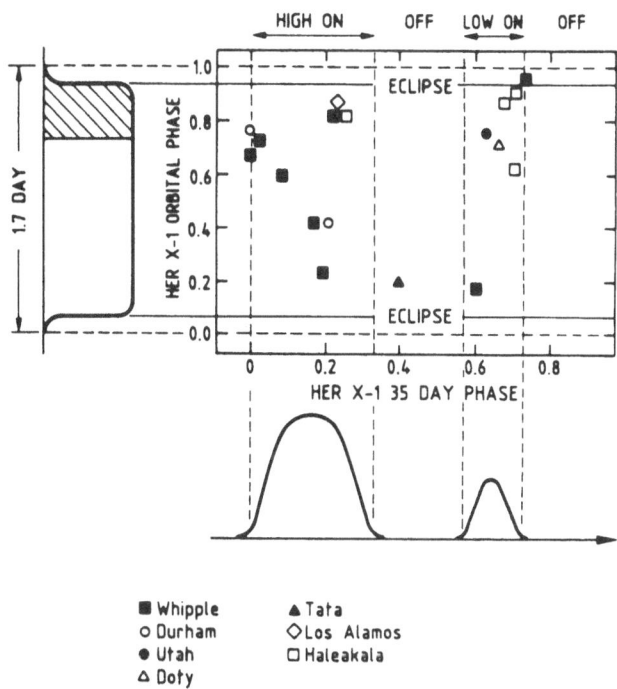

Fig. 9. Reported VHE/UHE Hercules X-1 detections versus 1.7 d and 35 d cycles (updated from Gorham 1986).

could be considered a solid proof for the emission of TeV and PeV photons from Her X-1: unfortunately when one looks at the details many clues do not add up, and we are once more left with the feeling that something tantalizing may be going on in the system without being able to construct a coherent picture for it.

The report by the Tata group (Vishwanath et al. 1989) stands out for its high significance: an excess of counts 42σ above the background was observed while their telescopes were tracking Her X-1. A problem with the data recording system prevented the Authors from identifying the pulsed signature in their data base and forced them to call their claim *possible*. The other four groups (Resvanis et al. 1988; Lamb et al. 1988; Dingus et al. 1988; Gupta et al. 1990b) were able to carry out a periodicity analysis on their data. The results are shown in Figure 10. All four groups observe a peak in the spectrum at a period shorter than the X-ray by 0.16%. If of Doppler origin this would correspond to a line-of-sight velocity of $490\,\mathrm{Km\,s^{-1}}$, more than a factor of 3 higher than the observed orbital velocity. Two groups, Whipple (Lamb et al. 1988) and Los Alamos (Dingus et al. 1988), had the possibility to distinguish the nature of the primary radiation: Whipple through the use of the imaging technique, Los Alamos by measuring the muon content of showers. Both failed to confirm the photonic nature of the primary radiation. As mentioned above for Cygnus X-3, this is hard to explain, especially at TeV energy where predictions based on the behaviour of photon induced EAS have allowed the Whipple group to detect the Crab Nebula. The Ooty claim (Gupta et al. 1990b) is based on 4 days (July 1, August 8-9, and November 21) in 1986 selected for their higher than average count rates. The four data sets were linked in phase and the Fourier analysis performed yielded a significant peak at the period of 1.2357701 s. Lewis, Lamb, and Biller (1990) have pointed out that this very precise determination of the period implies that the *clock* responsible for the modulation is one order of magnitude more stable than the neutron star in the system. They also argue that because of large gaps in the data the search in period is bound to yield a positive result irrespective of the behaviour of the source.

3.4 Vela X-1

Vela X-1 is an X-ray binary containing a massive companion ($\approx 24 M_\circ$) and a neutron star spinning with a period of 283 s; the orbital period of the system is 8.9 d.

It was first identified in the UHE range by the University of Adelaide Group (Protheroe, Clay, and Gerhardy 1984). Data collected from 1979 to 1981 were first subjected to an age cut ($s > 1.3$), then folded at the orbital period. With an excess of 11 counts seen at orbital phase 0.63 the flux was estimated to be $(9.3 \pm 3.4) \times 10^{-15}\,\mathrm{cm^{-2}\,s^{-1}}$ above 3×10^{15} eV (99.7% confidence level). Poor time recording prevented from looking at the pulsar. Tentative supporting eveidence for this detection came from the re-analysis of EAS data by the BASJE collaboration (Suga et al. 1985) and the Potchefstroom air shower array (van der Walt et al. 1987). It has to be remarked, however, that the emission is seen at different orbital phases.

The Potchefstroom group first reported evidence for VHE γ-ray emission from the system (North et al. 1987; Raubenheimer et al. 1989). Analysis of data taken in 1986-87 shows that Vela X-1 is a persistent source of γ rays modulated at the neutron star period. The flux above 2 TeV was measured in $(1.5 \pm 0.4) \times 10^{-11}\,\mathrm{cm^{-2}\,s^{-1}}$ (99.7% confidence level). A weak modulation of the signal with the orbital period was also observed. Besides this constant emission, the source shows episodes of

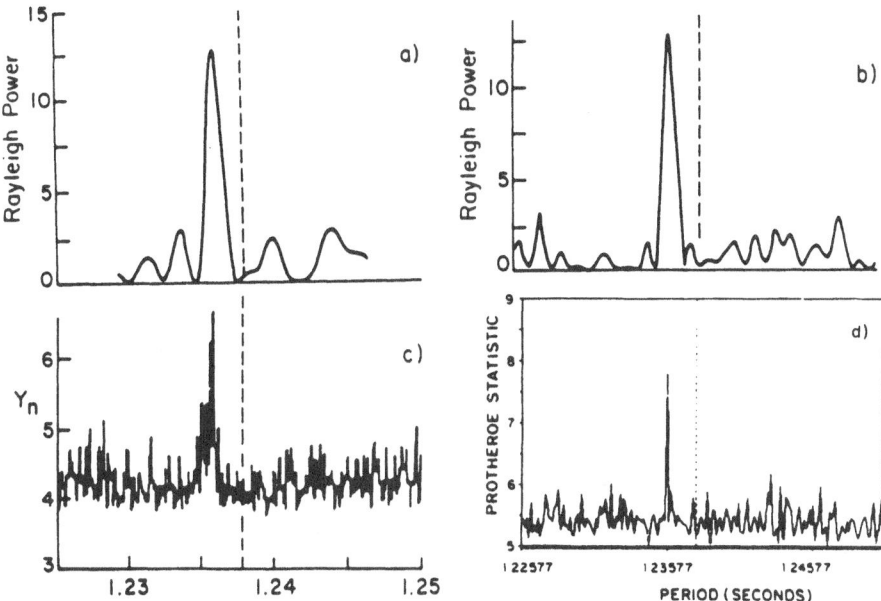

Fig. 10. Detection of Hercules X-1 at a blu-shifted frequency in 1986: (a) Resvanis et al. 1988; (b) Lamb et al. 1988; (c) Dingus et al. 1988; (d) Gupta et al. 1990.

increased activity lasting a few pulsar rotations. One of these happened after the neutron star had entered the X-ray eclipse. This behaviour, also seen on one occasion in Her X-1 (Gorham et al. 1986), suggests that γ rays are produced at a different site than X rays. VHE emission from Vela X-1 was confirmed by the Durham group (Carramiñana et al. 1989). Data were collected during 1986-88 with the telescope the group operates in Narrabri, South Australia. The emission appears to be modulated at the neutron star period. No evidence is found for episodes of increased emission lasting a few pulsar periods, nor for any modulation of the signal with the orbital motion. The reported flux is $(7.4 \pm 1.5) \times 10^{-11}\,\mathrm{cm}^{-2}\,\mathrm{s}^{-1}$ above 300 GeV, at a 99.99% confidence level.

Both the Potchefstroom and the Durham data show that the neutron star period changes unpredictably with time, undergoing phases of spin up and spin down. This behaviour is seen in X rays as well (Figure 11). The TeV observations seem to be in good agreement with other data.

4 Final Remarks

The Crab Nebula is firmly established as a source of TeV γ rays. Its emission is constant and several independent groups have confirmed its existence at comparable flux levels. Two experiments have established the signal is photonic in nature thank to the use of the imaging technique. No *ad hoc* hypotheses have to be made to explain the signal from the Nebula.

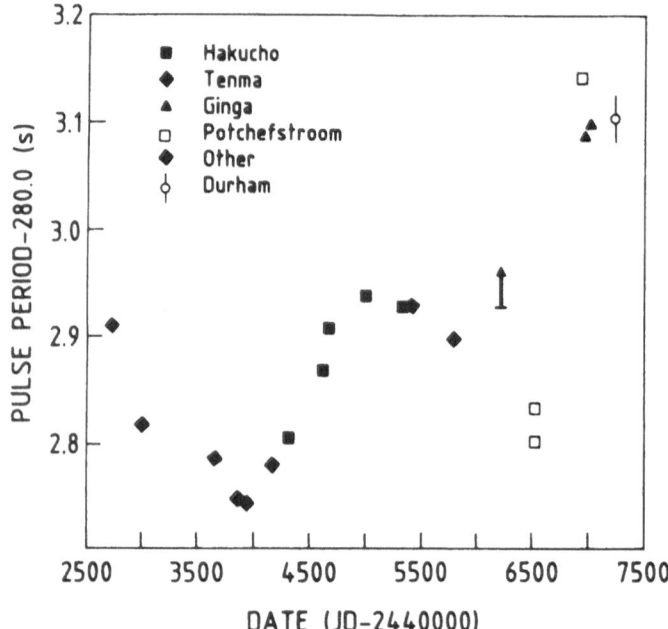

Fig. 11. Period history of Vela X-1. The VHE reports are shown with open symbols (updated from Raubenheimer et al. 1989).

The signals claimed from X-ray binaries do not enjoy the same consistency, as described in the previous chapters. In attempting an interpretation we are left with three hypotheses (and three questions).

a. The interaction of photons with matter changes drastically above 1 TeV becoming more hadron-like. Note however that in the case of the Crab Nebula the photonic nature of the signal has been established on the basis of EAS models who draw heavily upon standard physics, which therefore seems to be working fine up to a few TeV. This cannot be the reason why the Cherenkov imaging technique fails to reveal any signals from X-ray binaries. The nature of the interaction should then change somewhere between 10 TeV and 100 TeV. This, however, cannot be explained in the framework of QCD (Halzen et al. 1990). Does QCD stop working above 10 TeV? *Do theoreticians know what they are doing?*

b. A neutral particle, still unknown, is being produced by Cosmic Particle Accelerators. If so, its mass should be less than $\approx 2\,\mathrm{MeV}/c^2$ in order to preserve phase coherence of the signals. A particle in this mass range should already have been detected in some of the current accelerator experiments. *Do particle physicists know what they are doing?*

c. The signals we are seeing are just wicked statistical fluctuations we do not yet understand. This is certainly true for some of the reported cases. *Do γ-ray astronomers know what they are doing?*

In 1991 two more sensitive experiments will be coming on line: GRANITE and CASA. This, together with the long-awaited launch of GRO, will give us for the first time a 10 energy decades window on the Universe. Observations from these and already operational experiments will give us more puzzles and, hopefully, some answers. Watch for it, the best is still to come!!

5 Acknowledgements

I would like to thank the organizers of this meeting for their kind hospitality in Suhora. My thanks also go to V. Basiuk, P. Goret and P.T. Reynolds for discussions and help with the figures. T.C. Weekes is thanked for his words of wisdom.

This work is supported by the tax-payers of the European Economic Community.

References

Akerlof, C., et al. 1990 Nucl. Phys. B (Proc. Suppl.) 14A, 237

Akerlof, C., et al. 1990, Proc. 21st I.C.R.C. (Ed. R.J. Protheroe), 2, 135

Baltrusaitis, R.M., et al. 1985, Ap. J. (Letters), 293, L69

Baht, P.N., Ramana Murthy, P.V., and Visvanath, P.K. 1988, J. Atrophys. Astron. India, 9, 155

Baht, P.N., et al. 1990, Proc. 21st I.C.R.C. (Ed. R.J. Protheroe),2 , 148

Basiuk, V., et al. 1989, Proc. International Workshop 'Very High Energy Gamma-ray Astronomy', Crimea, USSR (Eds. A.A. Stepanian, D.J. Fegan, M.F. Cawley), p. 41

Basiuk, V., et al. 1991, Proc. International Conference on High Energy γ-ray Astronomy, October 2-5, 1990, Ann Arbor, Michigan, USA (in press)

Bonnet-Bidaud, J.M., and Chardin, G. 1988, Phys. Rep., 170, 325

Brazier, K.T.S., et al. 1990, Ap. J., 350, 745

Carramiñana, A., et al. 1989, Ap. J., 346, 967

Cassiday, G.L., et al. 1989, Phys. Rev. Letters, 62, 383

Cawley, M.F., et al. 1990, Exper. Astron., 1, 173

Chadwick, P.M., et al. 1985, Astron. Astrphys., 151, L1

Chadwick, P.M., et al. 1985, Nature, 318, 642

Chadwick, P.M., et al. 1985, Proc. 19th I.C.R.C., 1, 251

Chardin, G., and Gerbier, G. 1989, Astron. Astrophys., 210, 52

Ciampa, D., Clay, R.W., and Edwards, P.G. 1989, Ap. J., 346, 151

Dingus, B.L., et al. 1988, Phys. Rev. Letters, 61, 1906

Dowthwaite, J.C., et al. 1984a, Ap. J. (Letters), 286, L35

Dowthwaite, J.C., et al. 1984b, Nature, 309, 691

Dzikowski, V., et al. 1983, Proc 18th I.C.R.C., 2, 132

Fazio, G.G., et al. 1972, Ap. J. (Letters), 175, L117

Fegan, D.J., et al. 1989, Astron. Astrophys., 211, L1

Fegan, D.J. 1990, Proc. 21st I.C.R.C. (Ed. R.J. Protheroe), 11, 23

Gaisser, T.K. 1990, Science, 247, 1049

Gibbs, K.G., et al. 1988, Nucl. Inst. Meth., A264, 67

Goret, P., et al. 1988, Nucl. Inst. Meth., A270, 550

Gorham, D.W. 1986, Ph. D. thesis, U. of Hawaii (unpublished)

Gorham, D.W., et al. 1986, Ap. J. (Letters), 308, L11

Gregory, A.G., et al. 1990, Astron. Astrophys., 237, L5

Gupta, S.K., et al. 1990a, Proc. 21st I.C.R.C. (Ed. R.J. Protheroe), 2, 162

Gupta, S.K., et al. 1990b, Ap. J. (Letters), 354, L13

Halzen, F. et al. 1990, Proc 21st I.C.R.C. (Ed. R.J. Protheroe), 9, 142

Hillas, A.M. 1985, Proc. 19th I.C.R.C., 3, 445

Hillas, A.M. 1987, Proc. 20th I.C.R.C., 2, 362

Hillas, A.M., and Patterson, J.R. 1987, NATO ASI Series 'Very High Energy Gamma-ray Astronomy' (Ed. K.E. Turver), Vol. C199, p. 243

de Jager, O. C. 1987, Ph.D. Thesis, Potchefstroom University (unpublished)

de Jager, O. C., Raubenheimer, B. C., and Swanepoel, J. W. H. 1988, Proc. 3rd Workshop 'Data Analysis in Astronomy' June 20-27, Erice, Italy

Kirov, I. N., et al. 1985, Proc. 19th I.C.R.C., 1, 135

Koen, C. 1990, Ap. J., 348, 700

Kwok, P. 1990, private communication

Lamb, R.C., and Weekes, T.C. 1986, Astrophys. Lett., 25, 73

Lamb, R.C., et al. 1987, NATO ASI Workshop 'Very High Energy Gamma-ray Astronomy' (Ed. K.E. Turver), Vol. C199, p. 139

Lamb, R.C., et al. 1988, Ap. J. (Letters), 328, L13

Lang, M.J., et al. 1990, Proc. 21st I.C.R.C. (Ed. R.J. Protheroe), 2, 139

Lawrence, M.A., Prosser, D.C., and Watson, A.A. 1989, Phys. Rev. Letters, 63, 1121

Lewis, D.A. 1989, Astron. Astrophys., 219, 352

Lewis, D.A, Lamb, R.C., and Biller, S.D. 1991, to appear in Ap. J.

Macomb, D.J., et al. 1991, submitted to Ap. J.

Mayer-Hasselwander, H.A., and Simpson, G. 1990, Proc. 21st I.C.R.C. (Ed. R.J. Protheroe), 1, 261

Middletich, J., et al. 1985, Ap. J., 292, 267

Mihara, T., et al. 1990, Nature, 346, 250

Mukanov, J. B. 1983, Izv. Krym. Astrfiz. Obs., 67, 55

Nagle, D. E., Gaisser, T. K., and Protheroe, R.J. 1988, Ann. Rev. Nucl. Part. Sci., 38, 609

Nel, H.I., et al. 1990, submitted to Ap. J.

North, A.R., et al. 1987, Nature, 326, 567

Protheroe, R.J., Clay, R.W., and Gerhardy, P.R. 1984, Ap. J. (Letters), 280, L47

Raubenheimer, B.C., et al. 1986, Ap. J. (Letters), 307, L43

Raubenheimer, B.C., et al. 1989, Ap. J., 336, 394

Resvanis, L., et al. 1987, NATO ASI Series 'Very High Energy Gamma-ray Astronomy' (Ed. K.E. Turver), Vol. C199, p. 105

Resvanis, L.K., et al. 1988, Ap. J. (Letters), 328, L9

Samorsky, M., and Stamm, W. 1983a, AP. J. (Letters), 268, L17

Samorsky, M., and Stamm, W. 1983b, Proc. 18th I.C.R.C., 11, 244

Stanev, T., Gaisser, T. K., and Halzen, F. 1985, Phys. Rev. D, 32, 1244

Suga, K., et al. 1985, Proc. workshop 'Techniques in Ultra High Energy Gamma Ray Astronomy' (La Jolla), (Eds. Protheroe, R.J., Stephens, S.A.) p.48

Teshima, M., et al. 1990, Phys. Rev. Letters, 64, 1628

Tumer, O.T., et al. 1985, Proc. 19th I.C.R.C., 1, 139

Tumer, O.T., et al. 1990, Proc. 21st I.C.R.C. (Ed. R.J. Protheroe), 2, 155

Vishwanath, P.R., et al. 1989, Ap. J., 342, 489

Vladimirsky, B.M., Stepanian, A.A., and Fomin, V.P. 1973, 13th I.C.R.C., 1, 456

van der Walt, E.J., et al. 1987, Proc. 20th I.C.R.C., 1, 303

Watson, A.A. 1985, Proc 19th I.C.R.C., 9, 111

Weekes, T.C. 1988, Phys. Rep., 160, 1

Weekes, T.C., et al. 1989, Ap. J., 342, 379

Weekes, T. C. 1990 private communication

High Energy and Very High Energy Gamma-Rays from Electromagnetic Cascade Induced by Relativistic Neutrons in AGN

A. M. Atoyan

Yerevan Physics Institute
Alikhanian Brothers St. 2, 375036 Yerevan, Armenia,USSR

1 The Model

In the models of active galactic nuclei (hereafter AGN) that assume efficient acceleration of the relativistic protons (RPs) in the vicinity of supermassive black hole (BH) at radii $r \lesssim r_0 \sim 10$ (in units of the gravitational radius of the BH), a significant fraction ($\sim 30\%$) of the power injected in RPs is transferred to the relativistic neutrons (RNs), which then escape from the RP acceleration region producing secondary RPs due to the RN decay at distances up to the kpc scales (Sikora, Begelman, and Rudak 1989; Kirk and Mastichiadis 1989; Atoyan 1990a). In the framework of this model, we consider the problem of electromagnetic cascade initiated due to *in situ* cooling of the secondary RPs and developed in the spatially inhomogeneous field of the background soft photons (with the spectral energy density $w_0(r, \epsilon_0) \propto \phi_r \phi_0(\epsilon_0)$, where $\phi_r =$const at $r > r_{mz} \sim 10^2$-10^3 and $\phi_r \propto r^{-2}$ at $r > r_{mz}$). (Hereafter, such a cascade will be called the N-cascade, to distinguish from a cascade in the spatially homogeneous field of the background photons, hereafter the H-cascade.) The soft photon spectra of the form,

$$\phi_0(\epsilon_0) \propto \frac{\epsilon_0^{\alpha_{rm}}}{(\epsilon_0 + \epsilon_c)^{\alpha_{mz}+\alpha_{rm}}}, \tag{1}$$

in the energy range $10^{-10} < \epsilon_0 \equiv h\nu_0/m_e c^2 < 1$ are supposed. Here ϵ_c ($\sim 10^{-7}$) corresponds to characteristic low-frequency cutoff (or flattening) of the AGN spectra in the IR-submillimeter range. The power-law spectra of the RNs with the exponent $\alpha \sim 2$ are assumed.

The following processes are taken into account:
a) e^+e^- pair production in photon-photon collisions;
b) inverse Compton scattering of the relativistic electrons (REs) off the background soft photons;
c) synchrotron emission of the REs in an ambient magnetic field.

It is shown that both bremsstrahlung and the ionization energy losses of the REs as well as e^+e^- annihilation of relativistic positrons on thermal electrons of accretion plasma can be neglected.

The gamma-ray spectra resulting from the N-cascade developed in the region $10 < r < 10^9$ are calculated numerically (for the details see Atoyan 1990b).

2 The results

1. The main parameter (analogous to the compactness parameter ℓ of the H-cascades) defining the development of the electromagnetic N-cascade is the integral luminosity L_{mz} of the continuous radiation of AGN from the submillimeter up to the X-ray bands. The critical value of L_{mz}, above which the N-cascade development becomes important, is $L_{cr} = 1.5 \times 10^{43}$ erg/s.

2. In contrast with the gamma-ray spectra resulting from the familiar electromagnetic cascades in the spatially homogeneous field of the background soft photons (the H-cascades), for the RN-induced N-cascade there exists no threshold energy above which an exponential cutoff of the resulting gamma-ray spectra would occur (see Figs. 1, 2).

3. With the increasing L_{mz}, the spectral efficiency of the N-cascade, $e(> E) \equiv L_\gamma(> E)/\dot{W}_n$ (with $\dot{W}_n$ being the total power transported by the RNs from the RP acceleration region $r \leq r_0 \sim 10$), decreases significantly (see Table 1). Taking into account that the efficiency of the conversion of the accretion plasma rest mass to the RN energy may be high, $\eta_n \equiv \dot{W}_n/\dot{M}c^2 \sim 0.03$ (see Atoyan 1990a), we conclude that the model under consideration may provide very high values of the resulting efficiency $\eta_\gamma(> E) \equiv L_\gamma(> E)/\dot{M}c^2 = \eta_n e(> E)$ of the accretion plasma rest mass conversion into high energy (HE) and very high energy (VHE) gamma-rays escaping AGN: $\eta_\gamma(> 1\text{GeV}) \sim (10^{-2}\text{-}3 \times 10^{-3})$ and $\eta_\gamma(> 1\text{TeV}) \sim (2 \times 10^{-3}\text{-}6 \times 10^{-5})$ for $L_{mz} \sim (10^{44}\text{-}10^{46})$ erg/s, respectively.

4. At energies $E \sim (1\text{-}10)$ GeV the γ-ray spectra can be approximated as $L_\gamma(E) \propto E^{-\alpha_\gamma}$ with the exponent $\alpha_\gamma = \alpha - 1 + \delta$, where $\delta \sim 0.1$ and α (~ 2) is the power-law index of the RPs (as well as of the RNs) in the RP acceleration region. With the gamma-ray energy increasing towards the VHE range the gamma-ray spectra gradually steepen, the steepening being faster for the sources with higher L_{mz}. Therefore the model predicts that for observations in the VHE range more favorable are those AGN which: (i) have relatively low luminosities in the submillimeter-IR band; (ii) reveal significant fluxes of soft gamma-rays, $E \geq 1$ MeV. From this point of view the nearby Seyfert galaxies, such as Cen A (already observed as the VHE gamma-ray source) and NGC 4151, seem the most favorable candidates. For the most powerful objects (such as QSOs, e.g., 3C 273) the VHE gamma-ray fluxes expected are very low, $F_\gamma(\geq 1\text{TeV}) \leq 10^{-12}$ cm^{-2} s^{-1} (see Fig. 2).

5. At energies $E \sim 1$ GeV, the neutrino fluxes expected are only $\sim 2\text{-}3$ times higher than the gamma-ray ones. For the powerful sources this difference becomes essentially greater in the VHE range (see dash-dotted line in Fig. 2), so that the

Table 1. The partial contributions $e_i(> E)$ of the i-th cycle of the N-cascade into the resulting spectral efficiency $e(> E)$ at different luminosities L_{mx}. The model parameters are: $\alpha = 2.0$, $\alpha_{mx} = 1.2$, $\alpha_{rm} = 0$, $\epsilon_c = 10^{-7}$, $r_{mx} = 300$, $\dot{M}c^2/L_{Edd} = 1$.

L_{mx}		10^{44} erg/s		10^{45} erg/s		10^{46} erg/s	
$E >$		1 GeV	1 TeV	1 Gev	1 TeV	1 GeV	1 TeV
	i=0	9.50	25.0	1.37	7.86	0.11	0.91
	i=1	49.6	49.0	27.0	31.5	13.0	15.7
	i=2	31.3	21.7	43.6	41.0	41.8	44.5
e_i/e	i=3	8.22	3.94	22.1	16.6	33.1	30.4
(%)	i=4	1.19	0.38	5.17	2.76	10.3	7.51
	i=5	0.11	0.02	0.64	0.23	1.57	0.83
	i=6	–	–	0.06	0.01	0.13	0.05
$e = \sum_i e_i$		0.24	0.060	0.18	0.020	0.098	0.0022

neutrino observations by DUMAND with the sensitivity $F_\nu(> 1\text{TeV}) \sim 10^{-10}$ cm^{-2} s^{-1} may be more effective than the gamma-ray observations of these sources by existing Cherenkov telescopes.

A similar problem of the RN-induced electromagnetic cascade in AGN has also been recently considered by Mastichiadis and Protheroe (1990).

References

Atoyan, A. M. 1990a, preprint YERPHI-1268(54)-90.

Atoyan, A. M. 1990b, preprint YERPHI-1269(55)-90.

Kirk, J. G., and Mastichiadis, A. 1989, *Astr. Ap.*, **213**, 75.

Mastichiadis, A., and Protheroe, R. J. 1990, *Astr. Ap.*, **246**, 279.

Sikora, M., Begelman, M. C., and Rudak, B. 1989, *Ap. J. (Letters)*, **341**, L33.

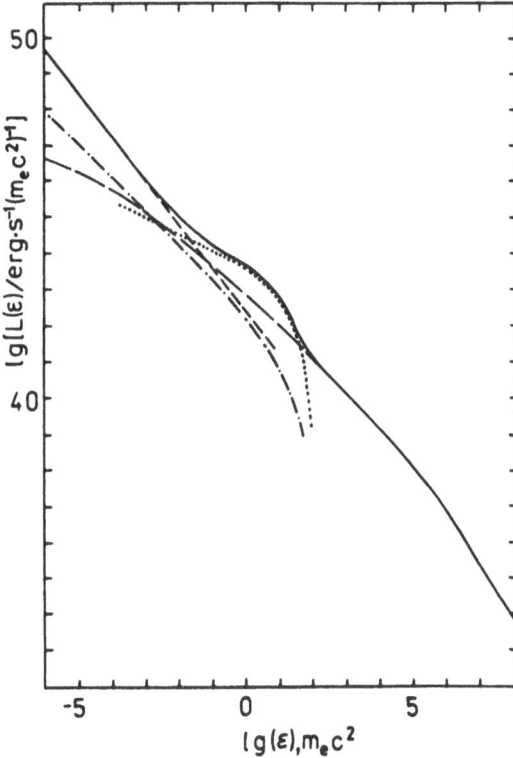

Fig. 1. The spectral luminosity $L(\epsilon)$ for AGN with $L_{mx} = 10^{45}$ erg/s, $\alpha_{rm} = 0$, $\epsilon_c = 10^{-7}$, $\alpha = 2$ ($\epsilon \equiv E/m_e c^2$).

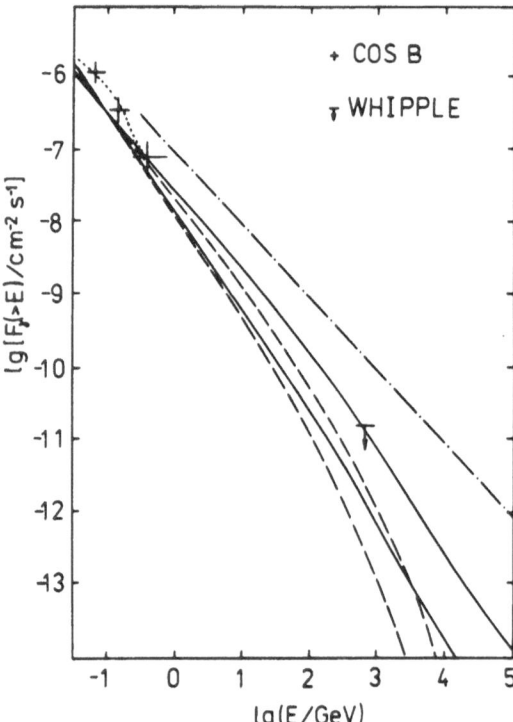

Fig. 2. The γ-ray fluxes expected from 3C273 (dashed curves) and NGC 1275 (solid curves) for $\alpha = 2.0$ and $\alpha = 2.3$. The observational data are presented for 3C 273.

Near Simultaneous Optical and TeV Observations of AE Aquarii

O.C. De Jager [1], P.J. Meintjes [1], B.C. Raubenheimer [1], C.

Brink [1], H.I. Nel [1], A.R. North [1], B. Visser [1], G. van Urk [1],

D. Buckley [2]

[1]Dept. of Physics, PU for CHE, Potchefstroom 2520, South Africa
[2]Dept. of Astronomy, Univ. of Cape Town, Rondebosh 7700, South Africa

Abstract: We present results of near simultaneous optical and TeV observations made on Sept. 14, 1990 of the novalike cataclysmic variable AE Aquarii. The optical was in a flaring state resulting in a bolometric luminosity which may have been high enough to give detectable TeV radiation during the TeV observation. During the latter we have seen periodic TeV emission at a frequency redshift of -0.7% relative to the spin frequency which confirms our earlier detection during 1989 at the 99.5% level. Both the optical and TeV did show evidence of a changing frequency with time which appears to be describing a single event. We also show that the shock above the white dwarf's polar cap may be collissionless so that protons may be accelerated to energies above 1 TeV. If the part of the accretion stream from the companion which reprocesses incident X-rays to optical pulsations is magnetized and dense enough, it may trap incident TeV particles for a long enough time to give TeV γ-rays through π^0 decay with the same period as in optical. The frequency will be constant if the spot of reprocessing is fixed in the binary frame, but shifted if the spot moves.

1 Introduction

The bright unusual novalike variable AE Aquarii ($m_v \approx 10.0 - 12.5$) is a binary system with an orbital period of 9.88 h consisting of a K5V red star and a white dwarf with a spin period of 33 s. The distance to this source is 84 pc (Bailey 1981) and the average accretion luminosity is $\approx 10^{33}$ erg.cm^{-2}.s^{-1} (Lamb and Patterson 1983). It shows ultraviolet flares and their risetimes are as short as 100 s (van Paradijs et al 1989). The bolometric luminosity thereof may increase up to 20 times the steady state value (Patterson 1979). Quasi periodic activity is seen at frequencies below the spin frequency, but QPO signals which mimic true periodic signals and redshifted by $\leq -1\%$ relative to the spin frequency have been seen when AE Aqr is in a flaring state (Patterson 1979). It also shows non-thermal radio flares on timescales of an hour or less (Bookbinder and Lamb 1988, Bastian, Dulk and Chanmugan 1988).

AE Aqr shows strong emission and absorption lines and the radial velocities of these two types of lines are antiphased by exactly 180°. Thus, the absorption lines follow the orbit of the K5V star whereas the emission lines are believed to follow the true orbit of the white dwarf (Robinson et al. 1991).

For an accretion rate of 10^{33} erg.s^{-1} Lamb and Patterson (1983) have inferred a surface field strength of $B \approx 6 \times 10^4$ G which allows centrifugal forces to be less than gravitational forces at the magnetospheric radius (which is close to the corotation radius) so that accretion may take place.

In this report we present results of TeV observations made on the same night when the optical flared. We confirm a previous TeV result and we concentrate at the similar results obtained in optical and TeV. We also show that a collissionless shock may accelerate protons up to TeV energies and that similar optical and TeV frequencies may result if the target material from the companion which reprocesses X-rays as pulsed optical light is also magnetized to trap TeV protons resulting in TeV γ-rays from proton-proton collisions.

2 The optical pulsations

Patterson (1979) discovered a 30.2327 mHz frequency showing two pulses spaced by half the period and one pulse stronger than the other. The semiamplitude of the Doppler delays is about 2.3 s. Patterson et al. (1980) also discovered pulsed X-rays with an intensity which is $\approx 20\%$ of the total X-ray flux from AE Aqr. It was also shown that the optical and X-ray pulse maxima coincide in phase. Further photometric observations (Robinson et al. 1991) of AE Aqr during 1982 and 1983 led to an improved optical frequency measurement of $\nu_{opt} = 30.2327088(2)$ mHz. The 60° phase shift between the Doppler orbit and the emission line orbit suggests that the pulse timing orbit describes the motion of the reprocessing region, and that the projected position of the region is uniquely determined to be at 2.3 light seconds from the center of mass of the binary system at an angle of 60° preceding the white dwarf. The reprocessing region may be some spot on the accretion stream completing its loop around the white dwarf, but the accretion disk associated with the white dwarf must then be thin enough or poorly defined not to absorb this stream. The radius to the reprocessing spot must be comparable to the Roche lobe radius r_L to give the observed pulse delays (Robinson et al 1991). The corresponding optical frequency is then an orbital beat frequency (Warner 1986)

$$\nu_{opt} = \nu_{spin} - \nu_{orb}$$

This property is seen from other intermediate polars. This implies that the true spin frequency of the white dwarf is probably $\nu_{spin} = 30.261$ mHz. The difference between ν_{spin} and ν_{opt} is small and it will be difficult to discriminate between them using the power spectrum from a single night.

3 Previous TeV observations of AE Aqr

Previous TeV observations by the Potchefstroom group (De Jager et al 1986) have been made between 1988 and 1989. One observation on July, 30 1989 have shown a strong periodic signal at a frequency of 30.03 mHz. Brink et al (1990) have shown that the period of this signal was not constant with time and have shifted towards the spin period on a timescale of two hours. Meintjes et al (1991) have shown that the frequency shift is comparable to, or, even smaller than the corresponding shifts seen in optical. Raubenheimer et al (1990) gave the result of a TeV pulse with marginal significance (obtained from all the 1988 and 1989 data) which is in phase with the optical and hence X-ray pulse maxima.

4 The optical and TeV survey of AE Aqr on Sept. 14, 1990

In an effort to arrive at a better understanding of the possible TeV emission from AE Aqr we have also made a few optical observations of this source with the 1 m reflector at SAAO during 1990. The observations were made in white light using a blue sensitive filter and mostly with a 1 s integration time. The first attempt at coordinated optical/TeV observations was made on Sept. 14. Two optical observations have been made: one immediately before (obs. S5302) and another (obs. S5304) immediately after TeV observation 67. In Fig.1 one can see from the optical that AE Aqr was in quiescence at the start of the TeV observation, but the average level of intensity must have increased towards the end thereof since a higher level of intensity is seen at the start of S5304. The optical indicated a 20 minute flare (around UT 22.7h) immediately after the TeV observation and remained in an enhanced state thereafter. Since the optical and TeV observations were not simultaneous, we have no way of telling whether AE Aqr did flare during the TeV observations. However, Sept. 14 did mark a time of flaring activity from AE Aqr.

A periodic TeV signal with a γ-ray luminosity of $\approx 3 \times 10^{32}$erg.s^{-1} will result in a detectable signal with a strength of about 10% of the cosmic ray background. This requirement for the luminosity is already large compared to the average accretion luminosity of $\approx 10^{33}$erg.s^{-1} given the conversion efficiency of acceleration power to γ-ray luminosity. However, the detection probability should increase when AE Aqr flares which may have been the case during the TeV observations of Sept. 14. It is difficult to estimate the bolometric correction to the flare luminosity since the flares have a strong ultraviolet component which may extend into the EUV.

TeV observation 67 started at UT 18.5853h, which is 7 minutes before the end of S5302, and ended 4 hours afterwards. We have binned the TeV counts in 20 minute intervals as shown in Fig. 1. The motivation for this bin width is to obtain

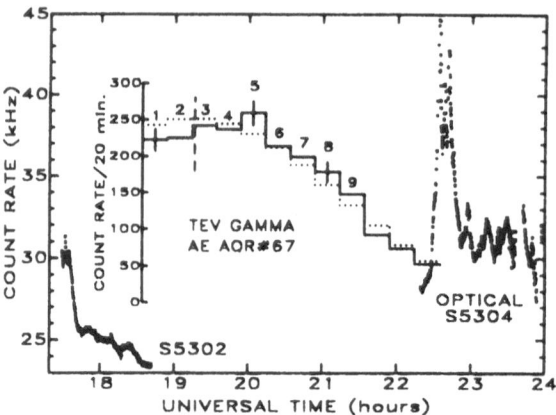

Fig. 1. The optical light curve (count rate in kHz) of AE Aquarii (SAAO observations S5302 and S5304 as indicated) on September 14, 1990. The data points shown are 10 s averages as a function of Universal Time (not corrected to the barycenter of the solar system). The count rate per 20 minutes for TeV observation 67 (solid line histogram) is also shown as a function of time. One standard deviation error bars are indicated on bins 1, 5 and 8. The vertical dashed line indicates the time of upper culmination and the dotted line histogram represents an estimate of the average count rate profile.

good statistics per bin for the sake of periodic analyses. It is interesting to note that the duration of the large optical flare is also 20 minutes, which may represent the typical timescale for TeV radiation during this night. The background count rate depends on the zenith angle z approximately as $N = N_0 cos^b(z)$ with $z = 26°$ degrees at upper culmination for AE Aqr (indicated on Fig. 1 by the vertical dashed line between bins 2 and 3). In the absence of simultaneous OFF-source observations we have performed a χ^2 goodness-of-fit on all the bins to obtain an average count rate profile. The minimum χ^2 value was 14.8 for 11 degrees of freedom, but there is a trend in the residuals – bins 5 to 9 are above the average level and the rest are below. This non-randomness could be the result of unknown atmospheric transmission effects since we did not have simultaneous OFF-source measurements, or, it may be due to a signal.

A periodogram of the whole TeV observation is shown in Fig.2 covering the frequency interval 28 to 32 mHz which includes the region of ≈ 28.8 to 30.2 mHz where strong quasi coherent periodic signals have been found previously (Patterson 1979). The strongest effect previously detected in TeV was at a frequency of 30.03 mHz as discussed in Section 3. This frequency is indicated by the vertical dotted line in Fig. 2. In observation 67 we see a peak at this same frequency which confirms our original detection at the 99.5% level after allowing for over-ampling in a ±1 independent Fourier spacing search interval around 30.03 mHz. The corresponding peakwidth is 0.37 mHz which implies a radiation timescale of $2/(0.37 \times 10^{-3}s) = 90$ minutes. The estimated signal strength is (10.5 ± 3.1)%, or, (225 ± 66) pulsed events. This effect may be visible as a DC-excess (with a duration of about 90 minutes) above the background estimate in Fig. 1 and the best candidate from the latter is the 100 min. interval in bins 5 to 9. A periodogram of this data subset gives a more significant peak (99.99% level) with the same width as the original detection (see Fig. 2). The signal strength here is (18.1 ± 4.5) % giving a total number of (181 ± 45) pulsed events. Since (1) the number of pulsed events in both data sets are statistical the same and (2) the widths of the two peaks on the periodogram are the same, they imply that we may have correctly identified the approximate interval of TeV radiation from Fig. 1 - despite the uncertainties in the atmospheric effects.

We have also tested the validity of this result by obtaining the power spectrum at other frequencies in this observation and many other observations taken from AE Aqr and of a background region taken after the AE Aqr observations and the results are consistent with white noise. Thus, there are no observable systematic requencies in the data. This will be reported elsewhere in detail.

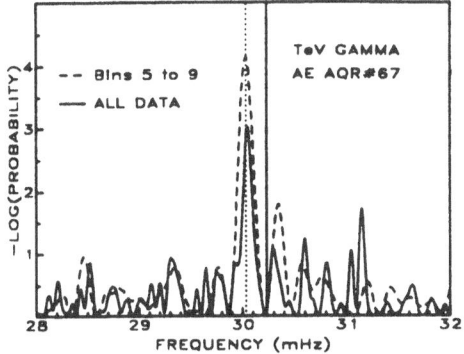

Fig. 2. The periodogram of the complete observation 67 (solid line) and bins 5 to 9 (dashed line) in terms of $-\log_{10}$ of the Rayleigh probability for chance occurence versus frequency. The vertical solid line indicates the stable optical frequency $\nu_{opt} = 30.2327\,\text{mHz}$ and the vertical dotted line represents the frequency of 30.03 mHz where a strong TeV signal had been previously found.

We have also calculated time evolved spectra for all the bins for which the statistics were good enough (i.e. 1 to 9). The results are shown in Fig. 3. The spectra of the first four bins show no significant structure, but the fifth bin shows a peak at a frequency of $(30.25 \pm 0.12)\,\text{mHz}$ (significance level: 99.94%) which is consistent with both ν_{opt} and ν_{spin}. Bin 8 also shows a significant structure at a redshifted frequency of $(29.75 \pm 0.14)\,\text{mHz}$. These two frequencies exclude each other at the 2.6σ level, but their net effect appears to be the cause of the 30.03 mHz signal in Fig. 2.

To see whether similar behaviour is seen in optical we have also calculated time evolved spectra for the optical data in data sets with window lengths ranging between 700 and 1200 seconds. This choice of duration arises from the need to have nearly similar count rates in each window. This reduces the effect of red noise in the power spectrum. To eliminate the effect of variable window lengths and count rates further we have calculated the signal strength (in a sinusoid) and plotted the square of this (giving a scaled Fourier power) versus frequency. The results for both optical observations are shown in Fig. 4. and the spectra may be directly compared with each other. It is clear that the signal in S5302 (and hence during the start of the TeV observation) was weak which is consistent with the non detection in TeV during the start of observation 67. The TeV signal appeared about 80 minutes after the start, but it is difficult to say whether the TeV activity persisted until after the start of S5304. The reason is that the sensitivity of the TeV telescope dropped towards the end of the observation due to the increasing zenith angle resulting in an increase in atmospheric absorption and threshold energy.

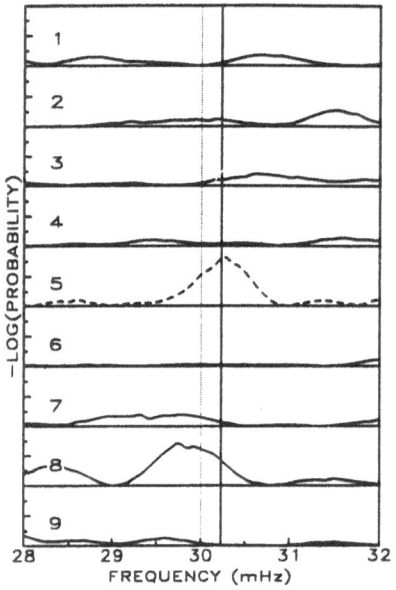

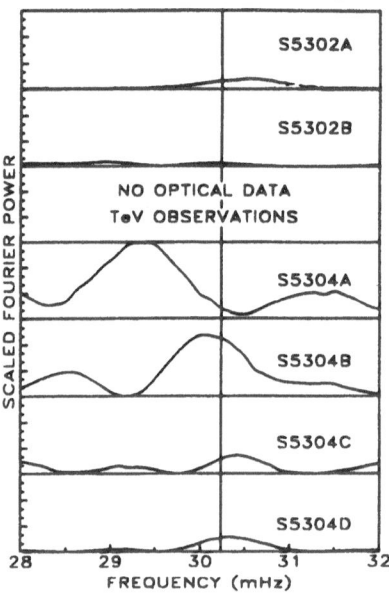

Fig. 3. The time evolved spectra for the bins marked 1 to 9 in the TeV count rate profile in Fig. 1. The vertical scale and the two vertical lines (solid and dotted) have the same meaning as in Fig. 2. The vertical scale consists of nine independent scales, each separated by a horisontal solid line, and ranging between zero and four with three tick marks in between.

Fig. 4. The time evolved spectra for optical observations S5302 and S5304. The power has been scaled as defined in the text and the six vertical divisions have the same range in power. They correspond to the following time intervals in UT (hours): S5302A: 18.028 - 18.361; S5302B: 18.361 - 18.694; S5304A: 22.316 - 22.511; S5304B: 22.511 - 22.775; S5304C 22.775 - 23.108; S5304D: 23.108 - 23.442.

From the optical and TeV power spectra we may be seeing a single event in time: The optical frequency started at ν_{opt} but with a low intensity towards the end of S5302. The TeV was also in quiescence during the first 80 minutes (i.e. bins 1 to 4 in Figs. 1 and 3), but was followed (bin 5) with a 20 minute burst with a frequency consistent with ν_{opt}. After this the TeV frequency changed to a redshift of $(-1.3 \pm 0.5)\%$ (i.e. bin 8, and probably bins 7 and 9 also). Optical observation S5304 started an hour later with a strong periodic signal and a frequency redshift of $\approx -2.8\%$. During the large optical flare the frequency shifted back to ν_{opt}.

5 Particle acceleration in AE Aqr

Bastian, Dulk and Chanmugan (1988) have modelled the radio flares of AE Aqr using the van der Laan model (VDL) as basis: A mildly relativistic electron spectrum is used and a single flare event is idealized as an impulsive injection of a spherical cloud of relativistic electrons of which the main energy loss mechanism is due to betatron losses when the cloud expands. The synchrotron losses in the magnetized cloud (initial $B \approx 25$ to $3000\,G$ for initial radii of 3×10^{10} to $10^9\,cm$ respectively) give rise to the radio spectrum with variable spectral index.

Cyg X-3 shows gaint radio flares with the VDL model as accepted basis (Bastian, Dulk and Chanmugan 1988) and the source of electrons may be due to bulk transport in jets which originates in a central engine. This interpretation is strengthened by the detection of jets in Cyg-3 (Spencer et al. 1986). The radio emission of AE Aqr appears to have been resolved at Arecibo (Lamb 1990) which may also be due to jets originating at a central engine. Fritz and Webb (1990) gave an analytical solution for the problem of electron acceleration in non-relativistic shocks due to impulsive injection. The result is an electron spectrum with a variable index which may explain the outburst behaviour of various radio sources. The optical flares from AE Aqr suggest that impulsive injection may be taking place in a non-relativistic shock (i.e. the upstream velocity is about 0.017c) giving bursts of accelerated electrons and hence the synchrotron emitting clouds.

In the above scenario the central engine is the result of a strong shock which is expected above the polar cap of the white dwarf. It was shown that accretion takes place in an accretion arc (i.e. in half an ellipse) on a fraction $f \approx 0.01$ of the stellar surface (see e.g. Lamb 1988). This gives a standing shock at a height which is much less than the stellar radius. A super-Alfvénic collisionless shock with Alfvén waves acting as scattering centers may develop if the free fall velocity exceeds the Alfvén velocity $V_A = B/\sqrt{4\pi\rho}$. This condition may be rewritten as a constraint on L giving (where $R \leq 10^9\,cm$ is the stellar radius and $B \geq 10^4\,G$ is the stellar magnetic field strength)

$$L > 3 \times 10^{32} B_4^2 R_9^{3/2} (M/M_\odot)^{1/2} \text{erg.s}^{-1}$$

For stronger magnetic fields this condition may still be met in the flaring state when $L \gg 10^{33}\text{erg.s}^{-1}$ is expected. Thus, we may expect the existence of a particle accelerator in AE Aqr which may only be active during the flaring state.

The maximum energy E_e^{max} which electrons may reach in a strong adiabatic shock is (Achterberg 1989)

$$E_e^{max} = 10^{10}(M/M_\odot)^{1/2} R_9^{-1/2} B_4^{-1/2} \text{ eV}$$

This cutoff is determined by synchrotron losses and it is unlikely that electrons are responsible for the TeV radiation, but such accelerated electrons may be responsible for the radio flares.

Synchrotron losses for protons and heavier nuclei are much less than for electrons and the limiting energy for protons (E_p^{max}) is about $10^{16}\,eV$ in the case of

AE Aqr if an infinitely large shock was available. However, the system's dimensions limits the maximum energy and we may obtain an estimate for the maximum energy (Harding 1990) by using the shock radius which is comparable to the stellar radius:

$$E_p^{max} = 5 \times 10^{13} B_4 R_9 \, \text{eV}$$

Thus, protons appear to be the appropriate candidates for very high energy γ-ray emission. The peak accretion luminosity from the white dwarf (confined to a small range of field lines) will then be L/f and may be as high as 10^{36} erg.s^{-1} during flares.

6 Gamma-ray production in AE Aqr

In the *keplerian frequency model (KFM)* the frequencies of the optical quasi periodic signals may be due to self luminous blobs rotating at keplerian speeds close to the corotation radius (Patterson 1979). If this model is to be applied to the TeV also, then the blobs must be emitting γ-rays isotropically and the periodicity is caused by the eclipse with the white dwarf: If the blob itself is magnetized then protons (from the white dwarf's polar cap) may be trapped for a sufficiently long time in the target so that their initial directions are isotropised to give TeV γ-rays through proton-proton collisions once the column density experienced by the proton inside the target exceeds ≈ 50 g.cm^{-2}. A TeV particle with energy E_p may be trapped by a magnetized blob if the Larmor radius of the particle is less than the radius s of the blob. This condition may be written as

$$B > 33(E_p/10\text{TeV})(s/10^9 \, \text{cm})^{-1} \, \text{G}.$$

This model allows the detection of redshifted and blueshifted frequencies for blobs rotating respectively outside and inside the corotation radius. However, the magnetic drag exerted by the white dwarf's field on the blob may result in a suppression of the blueshifted signals.

In the *beat frequency model (BFM)* for the TeV radiation the observed frequency redshift may arise from the trapping of TeV protons in the accretion stream from the companion star (provided that the Larmor radius is less than the radius s of the stream) at a radius r_γ from the white dwarf where the conditions in the accretion stream is appropriate for proton trapping and γ-ray production. The consequence of a trapped proton inside a moving target (with respect to the orbital reference frame) is the detection of a period shifted signal (Aharonian and Atoyan 1990). If the target moves in a keplerian orbit at a radius r_γ, we may expect a periodic signal at a beat frequency of

$$\nu_\gamma^s = \nu_{spin} - \sqrt{GM/r_\gamma^3}/(2\pi)$$

where $\sqrt{GM/r_\gamma^3}$ is the keplerian frequency of the target in question. This may be observable as a signal with a TeV frequency close to the spin frequency if r_γ is far removed from the corotation radius and a typical TeV frequency of 30 mHz (which corresponds to a frequency redshift of $\Delta\nu/\nu \approx -0.7\%$) will correspond to a radius of $r_\gamma \approx 4.3 \times 10^{10}$ cm which is still inside the Roche lobe radius ($r_L \approx 6.6 \times 10^{10}$ cm) of the white dwarf. If this same target is intercepted by the X-ray beam from the white dwarf, we may see the same frequency shift in optical (due to reprocessing) and possibly reflected X-rays (but at a lower intensity). Generally, a frequency redshift of $-0.3(r_\gamma/r_L)^{-3/2}\%$ is expected in the optical and TeV emission for such a target in a keplerian orbit at a radius $r_\gamma < r_L$. A neccesary condition for the KFM to work is that the trapping time (for γ-ray production to take place) should be *less* than the spin frequency of the white dwarf.

If the target in the orbit is fixed then we may expect reprocessed optical pulsed emission at a frequency ν_{opt} (as suggested by Robinson et al; see also Warner 1986). If this same spot is also magnetized having the minimum density for γ-ray production, then we may also expect the observation of TeV γ-rays with the same frequency as in the optical.

If the targets are not magnetised then we expect a much weaker TeV signal (due to the thin medium along the line-of-sight for proton-proton collisions to take place) at a frequency equal to ν_{spin}.

7 Conclusions

We have used the advantage of having optical and TeV observations of the novalike variable AE Aquarii on the same night and the presence of enhanced activity on the night of Sept. 14, 1990 indicated that it was in a high state. This eliminates a blind search through many nights' TeV data for a signal. This has the effect of an improvement in the sensitivity due to a decrease in the statistical penalties. However, this was also the first attempt at coordinated observations but we do not yet know what levels of optical activity will lead to detections.

The first detection of AE Aqr by Brink et al. (1990) was made without a detailed knowledge of the optical behaviour. Their basic result was the detection of a periodic signal in one previous observation with a frequency of 30.03 mHz which is redshifted by -0.7% relative to ν_{spin}. The basic result of this presentation is that we have *confirmed* the observation of Brink et al. at a statistical significance of 99.5% using all the data on Sept. 14. A more detailed analysis of the TeV data have shown that the periodic behaviour is similar to the optical behaviour with respect to the detection of a redshifted periodicity and a periodic signal consistent with either ν_{spin} or ν_{opt}, but most probably ν_{opt} from physical considerations. The optical and TeV frequency behaviour may be describing a single event with time.

A collissionles accretion shock may accelerate protons to energies > 1 TeV and we may expect the detection of pulsed TeV γ-rays at the same period as the optical (i.e. ν_{opt} and the QPO frequencies) if the targets which reprocess X-rays into optical are sufficiently magnetized so that protons are trapped for a long enough time to give γ-rays from proton- proton collissions. The BFM of Warner (1986) will then also apply to the TeV given favourable target conditions. However, from the results presented here it is difficult to discriminate between the BFM and the KFM. Finally, it would be interesting to see if the 30.03 mHz feature is a consistent feature in other TeV observations.

Acknowledgements

We would like to thank Darragh O'Donoghue for whipping up interest at UCT for optical observations of AE Aqr. We also thank A. Achterberg, A.M. Atoyan, R.A. Burger, H. Moraal, E. L. Robinson and B. Warner for useful discussions.

References

Aharonian, F.A., and Atoyan, A.M. 1990, preprint no. *YERPHI-1275(61)-90 (Yerevan, Armenia)*.

Achterberg, A. 1989, to appear in *NATO ASI Series C: Math. and Phys. Sci. Vol.* **305**, *Physical Processes in Hot Cosmic Plasmas*, ed. W. Brinkmann, A.C. Fabian, and F. Giovannelli (Kluwer acad. Publ.)

Bailey, J. 1981, *M.N.R.A.S*, **197**, 31.

Bastian, T.S., Dulk, G.A., and Chanmugan, G. 1988, *Ap.J.*, **324**, 431.

Bookbinder, J.A., and Lamb, D.Q. 1988, *Ap.J. (Letters)*, **323**, L131.

Brink, C., Cheng, K.S., De Jager, O.C., Meintjes, P.J., Nel, H.I., North, A.R., Raubenheimer, B.C., and van der Walt, D.J. 1990, in *Proc. 21st ICRC (Adelaide)*, **2**, p.283.

De Jager, H.I., De Jager, O.C., North, A.R., Raubenheimer, B.C., van der Walt, D.J., and van Urk, G. 1986, *South African J. Phys.* **9**, 107.

Fritz, K.D., and Webb, G.M. 1990, *Ap.J.*, **360**, 387.

Harding, A.K. 1990, *Nucl. Phys. B (Proc. Suppl.)*, **14A**, *3*.

Lamb, D.Q., and Patterson, J. 1983, in *IAU Colloquium 72, Cataclysmic Variables and Related Objects*, ed. M. Livio and G. Shaviv (Dordrecht: Reidel), p. 229.

Lamb, D.Q. 1988, in *Polarized Radiation of Circumstellar Origin*, ed. G.V. Coyne, S.J., A.M. Magalhães, A.F.J. Moffat, R.E. Schulte-Ladbeck, S. Tapia, D.T. Wickramasinghe (Vatican Observatory - Vatican City State), p. 151.

Lamb, D.Q. 1990, personal communication.

Meintjes, P.J., De Jager, O.C., Raubenheimer, B.C., O'Donoghue, D., Brink, C., Nel, H.I., North, A.R., van der Walt, D.J., van Urk, G., and Visser, B. 1990, to appear in *6th I.A.P. meeting - IAU Colloquium 129, Structure and emission properties of accretion disks*.

Patterson, J. 1979, *Ap.J.*, **234**, 978.

Patterson, J., Branch, D., Chincarini, G., and Robinson, E.L. 1980, *Ap.J. (Letters)*, **240**, L133.

Raubenheimer, B.C., North, A.R., De Jager, O.C., Meintjes, P.J., Brink, C., Nel, H.I., van Urk, G., and Visser, B. 1990, to appear in *High Energy Gamma-ray Astrophysics*.

Robinson, E.L., Shafter, A.W. and Balachandran, S. 1991, to appear in *Ap.J.*

Spencer, R.E., Swinny, R.W., Johnston, K.J., and Hjellming, R.M. 1986, *Ap.J.*, **309**, 694.

van Paradijs, J., Kraakman, H., and van Amerongen, S. 1989 *Astron. Astrophys. (Suppl. Ser.)*, **79**, 205.

Warner, B. 1986, *M.N.R.A.S.*, **219**, 347.

Lecture Notes in Physics

For information about Vols. 1–365
please contact your bookseller or Springer-Verlag

New Series m: Monographs